Military Strategy and the Origins of the First World War

AN *International Security* READER

EDITED BY
Steven E. Miller

PRINCETON UNIVERSITY PRESS
PRINCETON, NEW JERSEY

Contents

Introduction: Sarajevo After Seventy Years	Steven E. Miller	3
The First World and the International Power System	Paul Kennedy	7
Men Against Fire: Expectations of War in 1914	Michael Howard	41
The Cult of the Offensive and the Origins of the First World War	Stephen Van Evera	58
Civil-Military Relations and the Cult of the Offensive, 1914 and 1984	Jack Snyder	108
Windows of Opportunity: Do States Jump Through Them?	Richard Ned Lebow	147

Published by Princeton University Press, 41 William Street, Princeton, New Jersey 08540
In the United Kingdom: Princeton University Press, Guildford, Surrey

First Princeton Paperback printing, 1985
First hardcover printing, 1985
ISBN 0-691-07679-0 ISBN 0-691-02232-1 (pbk.) LCC 84-61326

Clothbound editions by Princeton University Press are printed on acid-free paper, and binding materials are chosen for strength and durability. Paperbacks, while satisfactory for personal collections, are not usually suitable for library rebinding.

The contents of this book were first published in *International Security,* a publication of The MIT Press under the sponsorship of The Center for Science and International Affairs at Harvard University.

The Great War and the Nuclear Age

Sarajevo After Seventy Years

Though distant in time, the disaster of 1914 continues to haunt the contemporary security debate. In the nuclear age, the images that remain from the summer of 1914—the escalation from an isolated event in a far corner of Europe to global war, the apparent loss of control of the situation by key decision-makers, the crowding out of diplomacy by military exigencies, the awful, protracted, often senseless slaughter on the battlefield—raise troubling doubts about our ability to forever conduct affairs of state safely in an international environment plagued by the ever-present risk of thermonuclear war. Hence the need to study anew, in the light of recent research and theoretical perspectives, the outbreak of war in 1914. The seventieth anniversary of the assassination of Archduke Ferdinand at Sarajevo provides an opportune moment to reflect on the origins and consequences of "The Great War" and to consider its lessons for our own time.

"Were we not," asked George Kennan in his own inquiry into the origins of World War I, The Decline of Bismarck's European Order, *"in the face of some monstrous miscalculation—some pervasive failure to read correctly the outward indicators on one's own situation . . . ? Must not the generation of 1914 have been the victim of certain massive misunderstandings, invisible, of course, to themselves but susceptible of identification today . . . ? Was there not a possibility that if we could see how they went wrong, if we could identify the tendencies of mass psychology that led them thus astray, we might see where the dangers lay for ourselves in our attempt to come to terms with some of the great problems of public policy of our own day?" In raising these questions, Kennan captures, with his customary eloquence, the underlying motivation for the present collection of essays. The possibility that there could be another "monstrous miscalculation" makes it imperative that we face the future with the fullest possible comprehension of the mistakes of the past, that we search the wreckage of that earlier accident for clues about its cause. For those who believe that the 1914 analogy has relevance for the nuclear age, these essays provide thoughtful reminders of the dangers to which it alludes.*

If there is a single message that leaps from these pages, it is that the wellspring of disaster in 1914 resided in the unwavering and wholehearted commitment of all the major European militaries to offensive doctrines—despite the overwhelming defensive advantages afforded by prevailing technologies. It was this profound belief in the primacy of the offensive that linked the mobilization plans of the major powers, one to another, that made speed necessary and preemption desirable, that limited the opportunities for diplomacy while creating an irresistible dynamic of escalation, that took policy from the hands of the political leaders and put it in those of the generals. By the summer of 1914, all the major continental powers—in particular, France,

Germany, and Russia—had adopted offensive military doctrines and in the July crisis this collective offensive-mindedness resulted in such severe political-military instability that a single murder in Serbia triggered a world war.

Three of the essays that follow focus on aspects of the cult of the offensive. In "Men Against Fire: Expectations of War in 1914," Michael Howard describes what *European militaries believed about the nature of warfare in the period leading up to the war. He documents their relentless adherence to doctrines which emphasized morale, calvary charges, bayonet assaults, and offensive spirit at a time when the emergence of rapid-fire weapons, barbed wire, and entrenchments were making such operational inclinations not merely obsolete but suicidal. He provides evidence as well of their willful neglect of indications that their preferred doctrines had been rendered ineffective by new technologies and tactics. The result was offensive doctrines, completely at odds with prevailing military reality, that not only contributed to the outbreak of war but spent tens and hundreds of thousands of lives on mere yards and miles of strategically meaningless turf.*

Jack Snyder, in "Civil-Military Relations and the Cult of the Offensive," seeks to explain why *European militaries preferred offensive doctrines as they did. He argues that the cause of this intense pursuit of offense in Germany, France, and Russia was crises of civil-military relations which led militaries to choose doctrines which enhanced their autonomy and minimized the possibilities for civilian interference, that improved their ability to plan, that increased their importance in national policy, and that created the need for larger and better-funded military forces. By allowing militaries to plan on assuming the initiative and by making conquest seem feasible if not easy and external threats consequently seem great, offensive doctrines served these purposes even if, on the battlefield, they proved to be disastrously unsuccessful.*

Stephen Van Evera analyzes the consequences of Europe's passion for offense in "The Cult of the Offensive and the Origins of the First World War." These were, he suggests, almost without exception malignant. Belief in the power of the offensive made expansionist policies seem feasible and attractive even if only to make offensive operations by one's opponent more difficult. It raised preemptive and preventive war vulnerabilities and opportunities, and placed a premium on secrecy such that civilians were not fully aware of the plans of their own military nor cognizant that some of the dangers of diplomatic crisis remained hidden. It led also to the creation of ambitious and inflexible war plans, intended to provide a knockout blow, that made speed essential, the crossing of frontiers necessary even as a defensive precaution, and early mobilization imperative. Given the offensive nature of the war plans, mobilization meant war, while failure to mobilize early and quickly could be (or was thought to be) disastrous. All of these factors combined to put great pressure on diplomats, made crises difficult to control, and made diplomatic mistakes both extremely dangerous and difficult to correct. Applying these lessons to the nuclear age, Van Evera argues that Soviet and American decision-makers ought to be far more sensitive to the potential dangers of nuclear counterforce doctrines, which are essentially offensive.

Accompanying these three essays on the cult of the offensive are two that address the wider impact and implications of the First World War. Paul Kennedy examines the origins, the outcome, and the effects of the war in the context of the distribution of economic and military power among the major participants. He argues that the aggregate economic resources of the competing coalitions were a major factor in determining the course and result of the war and, further, that the war had surprisingly little impact on the underlying trends in the balance of power. Kennedy indicates that the relative economic decline of the United States, when viewed in light of these findings, is an extremely damaging blow to its power and could eventually undermine its far-flung global position.

Finally, Richard Ned Lebow compares the 1914 case to more recent periods in Soviet–American relations in which a window of opportunity (or vulnerability) was thought to exist. He argues that there exist powerful constraints that make decision-makers reticent about choosing war even when they face a window of opportunity—and that the effects of these constraints can be seen even in the 1914 crisis. This suggests, Lebow believes, that the existence of "windows" may be less dangerous than is suggested by the 1914 case, and by the recent alarm over a Soviet window of opportunity provided by the U.S. ICBM vulnerability problem, because states generally have strong reasons for not jumping through them.

The 1914 analogy is one frequently employed to help organize our thinking about the nuclear danger. The essays which follow offer interpretations of the proper meaning of that analogy in the context of the nuclear age. Barbara Tuchman has written that World War I "lies like a band of scorched earth" across modern history, dividing what came before with what came after. That this would be incalculably more true of nuclear war justifies our every effort to find wisdom in the lessons of that earlier catastrophe.

—The Editors

The First World War and the International Power System

Paul M. Kennedy

This essay examines the First World War in the context of what has been called "the shifting balance of world forces."[1] It concerns itself with the grand strategy of the powers in the largest sense of that term, that is, the position occupied by each of the major combatants within the global order and the extent to which that position affected, and was affected by, the war. In consequence, it will have little to say upon the day-by-day diplomacy of the pre-1914 era, the domestic-political background to the war, public attitudes towards interstate conflict, and the detailed operational plans of the various states.[2] By contrast, the geopolitical and economic alterations of the age will be given considerable attention, since it will be argued that they not only illustrate the peculiar linkages in the power constellation by 1914, but also help to explain the general military equilibrium, and the ultimate *outcome,* of the prolonged conflict itself.

That the outbreak of war was preceded by an arms race of staggering proportions is well known. As the following statistics show, *every* power was devoting far greater funds to military expenditures by 1914 than had been the case two or three decades earlier. While this was true even of more distant states like the U.S. (following the Spanish–American War) and Japan (following the Russo–Japanese War), the center of the arms race was clearly in Europe. Between 1900 and 1914, military expenditures more than doubled

---

Paul M. Kennedy is the J. Richardson Dilworth Professor of History at Yale University.

---

1. Charles L. Mowat, ed., *The New Cambridge Modern History,* Vol. 12, *The Shifting Balance of World Forces, 1898–1945* (Cambridge: Cambridge University Press, 1968). For other relevant works, see Geoffrey Barraclough, *An Introduction to Contemporary History* (Harmondsworth, Middlesex: Penguin, 1967), chapters 3 and 4; A.W. DePorte, *Europe between the Superpowers: The Enduring Balance* (New Haven: Yale University Press, 1979); Paul Kennedy, "Mahan *versus* Mackinder," in idem, *Strategy and Diplomacy, 1870–1945* (London and Boston: Allen & Unwin, 1984).
2. On these issues one can consult, *inter alia,* Luigi Albertini, *The Origins of the War of 1914,* trans. and ed. Isabella M. Massey, 3 vols. (London: Oxford University Press, 1952–57; reprint ed., Westport, Conn.: Greenwood Press, 1980); A.J.P. Taylor, *The Struggle for Mastery in Europe, 1848–1918* (Oxford: Clarendon Press, 1954); Barbara W. Tuchman, *The Proud Tower: A Portrait of the World Before the War, 1890–1914* (New York: Macmillan, 1966); Arno J. Mayer, *The Persistence of the Old Regime: Europe to the Great War* (London and New York: Pantheon, 1981); L.C.F. Turner, *Origins of the First World War* (London: Edward Arnold, 1970); Paul M. Kennedy, ed., *The War Plans of the Great Powers, 1880–1914* (London and Boston: Allen & Unwin, 1979).

International Security, Summer 1984 (Vol. 9, No. 1) 0162-2889/84/010007-34 $02.50/0

in Russia, Germany, and Austria–Hungary and almost doubled in Italy, with the largest rises occurring after 1910 (see Table 1).

It is interesting to note that the expansion in appropriations was not accompanied by similar increases in regular military and naval personnel: the additional funds were primarily being devoted towards improved equipment, new field-guns, and larger warships (all expensive items) rather than upon infantrymen *per se,* although the latter did increase to some extent (see Table 2). Still, this issue can be a deceptive one, since all of the continental European states had large numbers of trained reserves, which could be mobilized in wartime to double or treble the size of the army. Overall national population—whether growing swiftly as in Russia and Germany, or infuriatingly static as in France—remained a significant military factor here, as it had been in previous periods of great-power conflict (see Table 3). But it was not the most significant factor, by any means.

Economic Measures of National Power

The problem with relying upon such simple statistical comparisons as the military/naval expenditures and personnel of the Great Powers is that they ignore the vital fact that modern warfare, if lasting more than a few months, would be hideously expensive and would therefore require the mobilization of national industrial and technological resources on an unprecedented scale. Should the war between the rival alliances not be "over by Christmas" (to use the happy phrase of August 1914), its outcome would depend more and more upon the *economic exploitation of the available industrial and agricultural bases of the two coalitions.*[3] The significance of the economic factor was not

Table 1. Defense Appropriations of the Powers ($ millions)

	Britain	France	Russia	Germany	Italy	Austria–Hungary	U.S.	Japan
1890	157	142	145	121	79	64	67	24
1900	253	139	204	168	78	68	191	69
1910	340	188	312	204	122	87	279	84
1914	384	197	441	442	141	182	314	96

Source: Quincy Wright, *A Study of War,* 2nd ed. (Chicago and London: University of Chicago Press, 1965), pp. 670–671.

3. This phrase is chosen to allow for the fact that Britain and France could secure additional resources from overseas (especially their empires and the United States); and could in turn

Table 2. Military and Naval Personnel of the Powers

	Britain	France	Russia	Germany	Italy	Austria–Hungary	U.S.	Japan
1890	420,000	542,000	677,000	504,000	284,000	346,000	39,000	84,000
1900	624,000	715,000	1,119,000	524,000	255,000	385,000	96,000	234,000
1910	576,000	769,000	1,225,000	694,000	322,000	425,000	127,000	271,000
1914	532,000	910,000	1,300,000	891,000	345,000	444,000	164,000	306,000

Source: Quincy Wright, *A Study of War,* 2nd ed. (Chicago and London: University of Chicago Press), pp. 670–671.

Table 3. Total Population of the Powers (in millions)

	Britain	France	Russia	Germany	Italy	Austria–Hungary	U.S.	Japan
1890	38	38	110	49	30	41	63	40
1900	41	39	133	56	32	45	75	45
1910	45	39	163	64	35	49	92	51
1914	45	39	171	65	37	52	98	55

Source: Quincy Wright, *A Study of War,* 2nd ed. (Chicago and London: University of Chicago Press, 1965), pp. 670–671.

totally ignored, of course, by the prewar planning staffs; and most of them took measures to ensure that gold stocks would be built up, and that civilian (bulk food and raw materials) usage of the railway networks would not come to a standstill because of mobilization. But all those were *short-term* expedients, to overcome the effects of what was planned to be a temporary dislocation. When a writer such as Ivan Bloch suggested that modern war between the Great Powers would be a lengthy, mutually exhausting affair, his arguments were derided by the military—or they were twisted around to sustain the claim that a swift victory *had* to be planned for, since the economy could not sustain a lengthy disruption of "normal" trade and industry.[4]

This pre-1914 assumption that a war fought between the powers would be decisive and quick was flawed in several crucial respects. The first was the military men's mistaken belief that offensive warfare would prevail over the foe's defenses—a strategical error, and a *mentalité,* which has been examined in many studies and is further analyzed in some of the other essays in this issue. The error was equally pronounced among the Admiralties, all of whom looked forward to a decisive battlefleet encounter and did not properly appreciate that the geographical contours of the North Sea and the Mediterranean and the newer weapons of the mine, torpedo, and submarine would make fleet operations in the traditional style very difficult indeed.[5]

theoretically transfer munitions to their Russian ally, just as Germany could aid its economically less advanced partners, if it so wished.

4. Kennedy, ed., *War Plans of the Great Powers,* pp. 18–19 and passim; Gerd Hardach, *The First World War, 1914–1918* (Berkeley and Los Angeles: University of California Press, 1977), p. 55 ff.; and, for the reaction to Bloch, see the comments in Roger Chickering, *Imperial Germany and a World Without War: The Peace Movement and German Society, 1892–1914* (Princeton, N.J.: Princeton University Press, 1975), p. 388 ff.

5. See the other essays in this issue; and, for naval developments, Paul M. Kennedy, *The Rise and Fall of British Naval Mastery* (London: Allen Layne, 1976), p. 199.

The second major flaw seems equally remarkable in retrospect: it lay in the alliance system itself.[6] The very existence of these coalitions meant that, even if a power was heavily beaten in a campaign and/or could see that its own resources were inadequate to sustain further conflict, it was encouraged to remain in the war by the hope—and promises—of aid from its allies. Austria–Hungary's dependence upon Germany is perhaps the clearest case of this; indeed, the alliance with Berlin not only kept Vienna in the war, it also prevented the Habsburg Empire after 1916 from getting out of it! On the Allied side, too, the agreement not to conclude a separate peace helped to keep each signatory in the conflict until (as in Russia's case) domestic collapse prevented any further prosecution of the war. Had Italy been alone after Caporetto, it would have left the war; and had France been alone after Verdun, it also might have been forced to sue for peace, as it did after Sedan some 46 years earlier. Thus, the alliance system itself virtually[7] guaranteed that the war would *not* be swiftly decided, and meant in turn that victory in this lengthy duel would finally go to the side whose combination of financial, industrial, and technological resources was the greatest.

For all these reasons, the only proper way we can begin to understand the relationship between the First World War and the global power system is to examine the indices of economic strength and efficiency before 1914. For the purposes of the argument which follows, it will also be useful to commence those indices in the Bismarckian period and to continue them into the interwar years. All these tables confirm the basic trends, but not quite in the same way or to exactly the same degree. Percentages of world manufacturing production, for example, include non-industrial (i.e., peasant-and-handicraft) production, and to that extent may emphasize unduly the position of the less developed among the Great Powers, such as Russia. On the other hand, steel production at this time was not only a measure of industrialization in its most modern form; it was also one of the key sinews of war. But perhaps the best measure of the extent of a nation's industrialization is its energy consumption from modern forms (that is, coal, petroleum, natural gas, and hydroelectricity, but not wood). Finally, the table of total industrial produc-

---

6. For a development of this argument, see L.L. Farrar, Jr., *The Short-War Illusion* (Santa Barbara and Oxford: Clio, 1973).
7. A sweeping statement. Of course there were elements of patriotism, hatred of the foe, and fear of domestic unrest, which prevented Russian, Austrian, Italian, and French politicians from negotiating a separate peace. But it is still difficult to believe they would have gone on fighting after a great defeat had they not been sustained—and constrained—by the alliances.

tion *relative* to the United Kingdom's total in 1900 reveals the pace as well as the extent of industrialization among the powers (see Tables 4–7). While admitting that, to some extent, all economic statistics are crude and even arbitrary forms of measuring *national power*,[8] it nonetheless seems clear that these tables point to significant conclusions which may be drawn about each nation and about the system as a whole.

JAPAN

For example, Japan was a marginal power before 1914, not only geographically but also industrially; in its steel production and energy consumption it was, for example, far behind even Austria–Hungary. At great cost to itself, it had built up a large navy and a considerable army, both well trained and willing to fight and die in the *samurai* tradition. It was no doubt fortunate that its two successful wars in this period were fought against an even more backward China (1894–95) and against a Czarist Russia (1904–05) which was militarily top-heavy and disadvantaged by the immense distance between St. Petersburg and the Far East. Nonetheless, Japan had required large-scale loans from the British and American money markets in order to sustain the

Table 4. Percentages of World Manufacturing Production, 1880–1938

	1880	1900	1913	1928	1938
Britain	22.9	18.5	13.6	9.9	10.7
France	7.8	6.8	6.1	6.0	4.4
Russia	7.6	8.8	8.2	5.3	9.0
Germany	8.5	13.2	14.8	11.6	12.7
Austria–Hungary (until 1914)	4.4	4.7	4.4	–	–
Italy	2.5	2.5	2.4	2.7	2.8
Japan	2.4	2.4	2.7	5.3	5.2
U.S.	14.7	23.6	32.0	39.3	31.4

Source: Paul Bairoch, "International Industrialization Levels from 1750 to 1980," *Journal of European Economic History,* Vol. 11, No. 2 (Spring 1982), p. 297.

8. Especially when used in the "concrete" form in L.L. Farrar, Jr.'s *Arrogance and Anxiety: The Ambivalence of German Power, 1848–1914* (Iowa City: University of Iowa Press, 1981), chapters 1 and 2, footnotes.

Table 5. Iron/Steel Production of the Powers, 1880–1938 (million tons) (pig-iron production for 1880 and 1890; steel production thereafter)

	1880	1890	1900	1910	1913	1920	1930	1938
Britain	7.8	8.0	5	6.5	7.7	9.2	7.4	10.5
France	1.7	1.9	1.5	3.4	4.6	2.7	9.4	6.1
Russia	0.47	0.95	2.2	3.5	4.8	0.16	5.7	18
Germany	2.4	4.1	6.3	13.6	17.6	7.6	11.3	23.2
Austria–Hungary (until 1914)	0.46	0.97	1.1	2.1	2.6	–	–	–
Italy	0.02	0.01	0.11	0.73	0.93	0.73	1.7	2.3
Japan	0.00	0.02	–	0.16	0.25	0.84	2.3	7
U.S.	3.9	9.3	10.3	26.5	31.8	42.3	41.3	28.8

Source: Figures taken from the "Correlates of War" print-out data made available through the Inter-University Consortium for Political and Social Research at the University of Michigan.

war against Russia.[9] Its strategical significance as Britain's partner in 1914 did not lie as a provider of economic resources (although some Allied loans were to be floated in Tokyo), but as an extra-European military force which could be thrown into the balance to eliminate the German presence in China and the Pacific and to protect British imperial interests east of Suez—precisely the opposite, in fact, of Japan's disturbingly anti-British and pro-German role in the 1930s.[10]

ITALY

Italy, too, was an economic lightweight; and the figures here confirm Taylor's observation that it "hardly climbed into" the ranks of the Great Powers.[11] At the outset of the First World War, it had not achieved even *one-quarter* of the industrial strength which Britain possessed in 1900; its steel production was

9. A.J. Sherman, "German–Jewish Bankers in World Politics: The Financing of the Russo–Japanese War," *Leo Baeck Institute: Yearbook XXVIII* (1983), pp. 59–73.
10. On which there is now a vast amount of recent literature, including: Paul Haggie, *Britannia at Bay: The Defence of the British Empire Against Japan, 1931–1941* (Oxford: Oxford University Press, 1981); James Neidpath, *The Singapore Naval Base and the Defence of Britain's Eastern Empire, 1919–1941* (Oxford: Oxford University Press, 1981); William R. Louis, *British Strategy in the Far East, 1919–1939* (Oxford: Oxford University Press, 1971).
11. Taylor, *The Struggle for Mastery in Europe*, p. xxviii. And see the further analysis in R.J.B. Bosworth, *Italy, the Least of the Great Powers: Italian Foreign Policy before the First World War* (Cambridge: Cambridge University Press, 1979), chapter 1.

Table 6. Energy Consumption of the Great Powers, 1880–1938 (in million metric tons of coal equivalent)

	1880	1890	1900	1910	1913	1920	1930	1938
Britain	125	145	171	185	195	212	184	196
France	29	36	47.9	55	62.5	65	97.5	84
Russia	5.4	10.9	30	41	54	14.3	65	177
Germany	47	71	112	158	187	159	177	228
Austria–Hungary (until 1914)	11.3	19.7	29	40	49.4	–	–	–
Italy	1.8	4.5	5	9.6	11	14.3	24	27.8
Japan	0.7	4.6	4.6	15.4	23	34	55.8	96.5
U.S.	68	147	248	483	541	694	762	697

Source: Figures taken from the "Correlates of War" print-out data made available through the Inter-University Consortium for Political and Social Research at the University of Michigan.

Table 7. Total Industrial Potential of the Powers in Relative Perspective (U.K. in 1900 = 100)

	1880	1900	1913	1928	1938
Britain	73.3	100	127.2	135	181
France	25.1	36.8	57.3	82	74
Russia	24.5	47.5	76.6	72	152
Germany	27.4	71.2	137.7	158	214
Austria–Hungary (until 1914)	14.0	25.6	40.7	–	–
Italy	8.1	13.6	22.5	37	46
Japan	7.6	13	25.1	45	88
U.S.	46.9	127.8	298.1	533	528

Source: Paul Bairoch, "International Industrialization Levels from 1750 to 1980," *Journal of European Economic History,* Vol. 11, No. 2 (Spring 1982), pp. 294, 299.

small; its energy consumption from modern sources was pathetic (and had not improved, relatively, by the eve of the next war). Its chief strategical function in European politics had been to act as a distraction, militarily, to Austria–Hungary (until Bismarck had cemented over those quarrels in the Triple Alliance of 1882); and then as a further distraction, both in military and naval terms, to France. The coming of the Anglo–French entente, with its awful implications for Italy's colonial, maritime, and economic future should there ever be a war between Triple Alliance and Triple Entente, accelerated Rome's tacit abandonment of its commitments to Berlin and Vienna. Its neutrality in 1914, and its entry into the war on the Allied side a year later, was no real surprise to seasoned observers. What *was* disappointing was the swiftness with which it was calling for Anglo–French aid, rather than providing new military and economic resources to the Allied cause. No doubt its plight would have been far worse had it been fighting the German army as its main opponent, instead of the companions of the good soldier Schweik. It was, to be sure, better to have Italy as a partner than as a foe; but the margin of benefit was not great. And the vast expansion of its armed forces could not be properly sustained on its own miniscule industrial base.

AUSTRIA–HUNGARY

Measuring Austria–Hungary's power before 1914 is a more difficult task. In steel production, energy consumption, and share of world manufacturing production, it was some way ahead of Japan and Italy, and not too far behind France; industrialization was under way, and at a quickening pace.[12] On the other hand, its ostensibly impressive population of 52 million (1913) concealed enormous ethnic diversities, a cumbersome dual monarchy, and substantial regional differences—all of which made it impossible for the Habsburg Empire to mobilize the manpower, and afford the military spending, which a smaller and much less populous France achieved. In alliance with Germany, Austria–Hungary formed a solid territorial bloc which it would be difficult to overwhelm, even from the east; and it was a source of surplus foodstuffs. While Vienna admitted that it could not take on the Russians alone, it could probably withstand a Franco–Russian (and even a Franco–Russian–Italian) coalition provided the axis with Berlin remained firm. Whether it could maintain its independence from Berlin if that Germanic victory occurred and whether, finally, it could sustain the pressures of a long, grinding war if the Anglo–Saxon powers were also involved were much more doubtful.

FRANCE

France, too, needed allies if it was to maintain itself against its much more powerful neighbor to the east. If the Battle of Sedan had destroyed the French army's primacy in west-central Europe, the following four decades of peace

---

12. For a brief survey, see N.T. Gross, "The Habsburg Monarchy, 1750–1914," in Carlo M. Cipolla, ed., *The Fontana Economic History of Europe*, Vol. 4, *Emergence of Industrial Societies*, Part 1 (London: Fontana, 1973).

had witnessed its industrial decline, at least *relative* to the faster-growing economies of Britain, Germany, Russia, and the United States. By the eve of war, its total industrial potential was only about 40 percent of Germany's, and its steel production little over a quarter. A further weakness was the curiously static nature of France's population, which in 1850 had been the largest in Europe (except Russia), and by 1913 had been overtaken by *all* of the other Great Powers. This meant that, to maintain itself against the large German army increases after 1912, France needed to conscript a far higher percentage (89 percent cf. 53 percent) of its eligible youth—with obvious economic consequences. As against these deficiencies, France had considerable strengths. Its accumulated wealth was substantial. It was ethnically homogeneous and, if attacked by Germany, would be politically united. The lessening of tensions with Italy meant that it was not geographically distracted to the south; and the friendship with Britain meant that it had access to the resources of the extra-European world—which became desperately important once the German armies overran many of France's centers of coal and iron/steel production. Nevertheless, what the latter point means is that France was *well placed to receive aid*; it certainly could not sustain itself long against Germany from its indigenous resources of manpower and material.

RUSSIA

In the decades before 1914, of course, it was to Russia rather than to Britain (let alone, the U.S.) that France looked for assistance against Germany. The actual power of the Czarist Empire was hard to assess at the time, and has not become any easier since. The vast growth in its population impressed all those who immediately translated that element into real military strength. Its army of over 1,300,000 in 1914 was far larger than any other, and backed by about 5,000,000 reserves—figures which made the younger Moltke sweat and French *revanchistes* crease with pleasure. Russia's military expenditures, too, were extremely high and, with the "extraordinary" capital grants on top of the fast-rising "normal" expenditures, may well have exceeded even Germany's total. Railway construction was proceeding at enormous speed—threatening within another few years to undermine the calculations upon which the Schlieffen Plan was based—and money was also being poured into a new Russian fleet. It is, moreover, worth noting that it was not merely the German and Austro-Hungarian General Staffs, looking to their eastern frontiers, which feared the Russian military colossus; there were also certain

Britons apprehensive about the impact of Czarist power in Asia. Russia, in the view of many observers, was the "coming" power.[13]

Despite all these signs of imposing military strength, however, Russia's economic backwardness was a stupendous burden. The crude figures alone show how far it had to catch up with the West before it became a *modern* industrial-strategical force in global affairs. In its steel production, energy consumption, and share of world manufacturing production, it was a considerable way behind both Germany and Britain; and when those figures are related to population size, and calculated on a per capita basis, the gap was a truly enormous one. In 1913 Russia's per capita level of industrialization was less than one-quarter of Germany's, and less than one-sixth of Britain's! In certain respects its swift industrial growth prior to the First World War was impressive, but this was concentrated in military-related, heavy industry sectors—steel, railways—to the detriment of the broader, more subtle areas of national strength (e.g., general levels of education, bureaucratic efficiency, technological expertise, experimental technology).[14] Even when resources *were* available, mental attitudes, educational backwardness, and the misjudgments attendant upon the general level of the Czarist decision-making and sociopolitical structure all combined to hamper national efficiency. The blunt fact was that, despite all the improvements since the setbacks of 1854–56, 1877, and 1905, the foundations of the Russian state and society were inadequate to bear the strains of a great-power war for very long. And while it might be true that the exigencies of the First World War increased industrial output (by the use, that is, of paper money and vastly expanded credit, which stimulated armaments production[15]), the strains were simply too great and the imbalances between military requirements on the one hand and economic and social stability on the other were too severe for a nation at Russia's particular stage of development.

---

13. The most thorough analysis is R. Ropponen, *Die Kraft Russlands* (Helsinki, 1968), which looks at Western perceptions of Russia's might before 1914; but see also the frequent coverage in Erwin Hölzle, *Die Selbstentmachtung Europas* (Göttingen, 1975).
14. For a new and very convincing survey of Russia's weaknesses as well as strengths, see D.C.B. Lieven, *Russia and the Origins of the First World War* (New York: St. Martin's Press, 1983), especially chapter 1.
15. Norman Stone, *The Eastern Front, 1914–1917* (New York: Charles Scribners, 1976), p. 287 ff. It is interesting to note that, in his latest book, *Europe Transformed 1878–1919* (London: Fontana, 1983), Stone admits that his earlier portrayal of Russia's military development before 1914 "overstates its case" (p. 425).

GERMANY

By comparison, the relationship between Germany's military power and its deeply based economic strength was much better balanced. Its population was second only to Russia's among the European states, and since Germans enjoyed far higher levels of education, social provision, and *per capita* income than Russians, the nation was strong both in the quantity and the quality of its population; in effect, it could sustain a very large regular army, and summon millions of reserves to the colors, with *less strain* than virtually every other combatant. Furthermore, although its military expenditures were enormous by 1914, they consumed less of the national income than those of any other state except Britain and the U.S.—that is to say, the enormous growth of the German economy was allowing it to bear the burden of armaments without undue discomfort. If, by the eve of war, its total energy production and its share of world trade had not yet overtaken Britain's, it was already ahead on most other measurements of national economic power. Its percentage of world manufacturing production was bigger, as was its total industrial potential. Its steel production was more than the British, French, and Russian output *combined*! Its lead in newer industries—chemicals and dyestuffs, electricals, machine tools, optics—was even more marked.[16] Behind this lay a broad-based infrastructure of good internal communications, a powerful and rich banking system, large numbers of engineers, technicians, and craftsmen, and impressive educational institutions.

It was true that this expansion of Germany's economy had led to a greater reliance upon maritime-based imports and exports, which in wartime would be vulnerable to a naval blockade; but given the country's technological expertise and its potential access to foodstuffs and raw materials in the East, that vulnerability was by no means as marked as Allied naval strategists fondly imagined.[17] Germany, in other words, had enormous staying power—always provided that the patriotism of the broad mass of workers, soldiers, and sailors outweighed their feelings of resentment that the governmental and social system unfairly favored the old Junker élites and big business. What weaknesses there were in the German system were *political* rather than

---

16. Stone, *Europe Transformed*, p. 159 ff.; Fritz Fischer, *War of Illusions: German Policies from 1911 to 1914*, trans. Marian Jackson (New York: W.W. Norton, 1975), part 1; W.G. Hoffmann, *Das Wachstum der Deutschen Wirtschaft seit der Mitte des 19.Jahrhunderts* (Berlin, 1965); W.O. Henderson, *The Rise of German Industrial Power, 1834–1914* (Berkeley and Los Angeles: University of California Press, 1976), part III.
17. See the discussion in Kennedy, *The Rise and Fall of British Naval Mastery*, p. 253 ff.

economic or military[18]—and even then, hardly as serious as the internal tensions of the Russian state.

The difficulties of either prewar bloc achieving a swift, "knock-out" blow in western or eastern Europe (because of the superiority of the strategical defensive and the existence of the alliance system) have been mentioned above. But a military deadlock would in turn swing the emphasis upon measuring national power from those initially mobilized divisions to the larger, industrial/technological advantages possessed by either side. On any of the criteria used in Tables 4–7 above, the inferiority of the German–Austrian coalition in terms of population and army size was massively compensated by a *clear superiority* in all other respects (see Table 8).

The figures in Table 8 do not include Italy, technically still on the side of the Central Powers before 1914; but its neutrality then, and even its switch to the other side in 1915, did not substantially alter the balances. If a lengthy war occurred, in which the industrial productivity and financial-cum-technological capacity of each side became even more important, then the *Franco–Russe* was clearly the weaker, and it is difficult to see how a German–Austrian victory could have been avoided without the intervention of *non*-continental European powers and their economic resources.

Table 8. Comparative Industrial/Technological Advantages

	Germany/ Austria–Hungary	France/Russia
Percentage of world manufacturing production (1913)	19.2	14.3
Energy consumption (1913), in metric million tons of coal equivalent	236.4	116.8
Steel production (1913), in million tons	20.2	9.4
Total industrial potential (U.K. in 1900 = 100)	178.4	133.9

18. Volker R. Berghahn, *Germany and the Approach of War in 1914* (London and New York: St. Martin's Press, 1974) is most useful here. This is not to say that there were *no* German military weaknesses—on this, see B.F. Schulte, *Die deutsche Armee* (Düsseldorf, 1977)—but they paled by comparison with the problems facing the Russian, French, and Austro–Hungarian general staffs.

BRITAIN

This was, to put it crudely, precisely the function which the British tradition of preserving the "balance of power" called for. While we can leave aside here any analysis of the heated contemporary debate upon whether the circumstances of 1914 demanded such an intervention,[19] an examination of Britain's own position and strengths is useful. In many of the more modern industrial spheres, as mentioned above, it had been overtaken by Germany; in older spheres, such as textile and coal production, and in shares of shipping (especially) and of world trade, it was still ahead. In 1914, it possessed a small, long-service army, which could only marginally affect the overall military balance—although that particular deficiency was made up by the numerical superiority of the Franco–Russian field armies. The real significance of Britain's intervention lay not in the short term (despite the acclaimed role of the British Expeditionary Force at the Marne), but in the middle-to-longer term.

In the first place, its Royal Navy could neutralize the threat presented by Germany's High Seas Fleet and steadily impose a commercial blockade upon the Central Powers—which is not to say that this would bring the latter to their knees, but that at least it would deny them access to sources of supply outside continental Europe. Conversely, it ensured completely free access to such sources for the Allied powers (except when later interrupted by the U-boat campaign); and this advantage was compounded by the fact that Britain was such a wealthy trading country, with extensive links across the globe and with enormous overseas investments, some of which at least could be liquidated to pay for dollar purchases. Diplomatically, this nexus of overseas ties—through treaties, trade, geographical position, cultural associations—meant that Britain's own decision to intervene influenced Japan's action in the Far East, Italy's declaration of neutrality (and later switch), and the benevolent stance, despite Woodrow Wilson's pleas for absolute neutrality,[20] of the U.S. Even more direct support was provided, naturally enough, by the self-governing Dominions and by India, whose troops swiftly moved into Germany's overseas empire and then against Turkey. All this massively

19. Zara S. Steiner, *Britain and the Origins of the First World War* (New York: St. Martin's Press, 1978), chapter 9; Paul Kennedy, *The Rise of the Anglo–German Antagonism, 1860–1914* (London and Boston: Allen & Unwin, 1980), p. 458 ff.
20. See the interesting analysis in John W. Coogan, *The End of Neutrality: The United States, Britain and Maritime Rights* (Ithaca, N.Y.: Cornell University Press, 1981), p. 176 ff.

outweighed the gains made by Germany in persuading smaller powers—Turkey itself, and Bulgaria—to join the Central Powers.

In addition, Britain's still-enormous industrial and financial resources could be deployed in Europe, both in raising loans and sending munitions to France, Belgium, Russia, and Italy, and in steadily building up a large army to be employed by General Haig in the campaigns along the Western Front. The economic indices again show the significance of Britain's intervention in power terms (see Table 9).

Given this remarkable *bouleversement* in the economic and strategical balances consequent upon Britain's intervention, the obvious questions arise as to why the Allies were failing to win the war even after two or three years of fighting—and by 1917 were indeed in some danger of losing it—and why they then felt it vital to secure American entry into the conflict. No definitive and empirically verifiable answer can be given to those questions, of course, but the following points, taken together, provide a plausible set of reasons. Despite the economic superiority of the Allies, that advantage was not so overwhelming as to lead to the swift overrunning of the Central Powers, particularly since: (a) each side possessed millions of troops who could not be defeated merely by one decisive battle in the style of Jena or Sadowa; (b) the tactical defensive, especially along the Western Front, usually worked to Germany's advantage—and implied that the Allies' numerical superiority would need to be even greater than it was to achieve a real breakthrough; (c) the geographical advantages of Germany's position, with good internal means of communication between east and west, to some degree compen-

Table 9. Industrial/Technological Comparisons with Addition of Britain

	Germany/ Austria–Hungary	France/Russia + Britain
Percentage of world manufacturing production (1913)	19.2	14.3 + 13.6 = 27.9
Energy consumption (1913), metric million tons of coal equivalent	236.4	116.8 + 195 = 311.8
Steel production (1913), in million tons	20.2	9.4 + 7.7 = 17.1
Total industrial potential (U.K. in 1900 = 100)	178.4	133.9 + 127.2 = 261.1

sated for its "encirclement" by the Allies, and, after Turkey had joined the war, helped to isolate Russia from its chief partners; finally, (d) the maritime blockade was much less effective against broad-based land powers than against island-states such as Britain or Japan, and it was further weakened by the German acquisitions and plunder in Luxemburg, northern France, and Rumania. Sea power could only work indirectly, and had to be applied in conjunction with the maximum military pressures on the Western, Eastern, and Italian fronts—putting the onus upon the Allies to mount ever larger and bloodier offensives.[21]

Finally, there was the question of *timing*. The relatively large resources which the Allies possessed could not be instantly mobilized in the pursuit of victory; indeed, it was not until 1916 that Haig's army totalled more than a million men, and even then the British were tempted to divert their troops into extra-European campaigns, thus reducing the potential pressure upon Germany.[22] This meant that, during the first two years of the conflict, Russia and France took the main burden of checking the *furor teutonicus*. Given the strains within Russian society, its steady isolation from its allies, and its military top-heaviness in relation to its small economic base, those years of fighting the German and Austro-Hungarian armies were more than enough. Even the 1916 Brusilov offensive, brilliantly executed against the unsuspecting Austrians, did virtually no damage to the major enemy, Germany, and placed yet more pressures upon Russian railways, foodstocks, and state finances. By the following spring, the position on the Eastern Front was irretrievable for Russia (and also pretty hopeless for Austria–Hungary), but excellent for Germany. Yet by that time the French army had also come close to collapse, after the slaughter around Verdun and the folly of Nivelle's offensive. What the presence of Haig's new armies did, therefore, was to make up for the increasing weariness of the French; they did *not* portend, in any strategically significant manner, an Allied victory in the west. In fact, Haig's penchant for offensives like Passchendaele squandered away these additional resources, and allowed Germany to continue to hold its own. In the south, too, Italy had shown a predictable incapacity to wage large-scale,

---

21. See the argument on these lines in Herbert W. Richmond, *National Policy and Naval Strength and Other Essays* (London: Longmans, 1928), pp. 71, 142; Kennedy, *The Rise and Fall of British Naval Mastery*, p. 253 ff.; both of which are confirmed in the economic analysis by Hardach, *The First World War*, p. 30 ff., which shows the rather limited effects of the naval blockade.
22. This is best covered in Paul Guinn, *British Strategy and Policy, 1914 to 1918* (Oxford: Oxford University Press, 1965).

industrialized warfare by itself and was calling for assistance. All this not only explains why the Allies had failed to achieve victory by 1917, but were eagerly looking towards the only power which seemed to have the resources to alter the depressing stalemate.

UNITED STATES

In so many ways, the disjunction between the economic strength and the military/naval strength of the United States was the obverse of that pertaining in Russia, Italy, and Austria–Hungary. By 1914 the American navy was a considerable force, capable of defending the western hemisphere; however, the relatively large expenditure upon the U.S. Army reflected higher wages and costs rather than size and fighting efficiency. Militarily, therefore, the United States was one of the least of the Great Powers. On the other hand, its economic growth during the preceding two or three decades was probably *the single most decisive shift in the long-term global power balances*. Between 1880 and 1913 the U.S. had not only eclipsed Britain as the world's leading industrial nation, but it was growing so fast that it was on the point of out-producing *all* of the European states combined! According to each of the Tables 4–7 above, its economic power in 1913 was larger than the next two countries (Germany and Britain) together, and its steel production and energy consumption were more than the next three countries (Germany, Britain, and Russia) combined. Since sweeping references to "the rise of the Superpowers" are frequently to be found in books upon modern world history, it is worth noting here the staggering *differences* between Russia and the U.S. by the eve of the First World War. The former possessed a front-line army about *ten* times as large as the latter's; but the U.S. produced *six* times as much steel, consumed *ten* times as much energy, and was *four* times larger in total industrial output (in *per capita* terms, it was *six* times more productive). While both countries were growing, they were certainly not growing in parallel.

Although some German advocates of *Weltpolitik* noticed these trends—and indeed used similar figures to argue that Germany needed to expand soon before the United States grew too powerful[23]—on the whole the implications of American industrial expansion were ignored by the general staffs and

23. Fischer, *War of Illusions*, p. 31 ff.; Holger H. Herwig, *Politics of Frustration: The United States in German Naval Planning, 1889–1941* (Boston: Little, Brown, 1976), part I.

politicians of continental Europe.[24] Sheer distance, the cozy assumption about the U.S. having "no entangling alliances," and the military planner's obsession about achieving a swift victory over neighboring foes, all help to explain this ignorance. Yet the older notion that the European leaders blindly committed national suicide by going to war in 1914, thereby undermining the continent's "primacy" in the world and giving it to the Americans, also needs to be considerably amended in the light of the statistics. According to one calculation, indeed, what the First World War did was to accelerate by a mere six years (from 1925, back to 1919) the time at which the U.S. would overtake Europe as the geographical area possessing the greatest economic output in the world.[25] Even had the First World War not happened, the "Vasco da Gama era"—the four centuries of European predominance over the rest of the globe—was swiftly coming to an end.

American intervention in the war could not immediately affect the *military* balances; even more than had been the case with Britain, vast organizational changes were necessary before a large-scale army could be landed in France. And while the addition of the U.S. Navy was welcomed, it did not alter the general strategical situation at sea. The collapse of the Russian army in 1917, and then of the entire Russian war effort, could not therefore be prevented—or even slowed down—by the fact of the American entry. Where the United States did make a vital contribution was on the *economic* front: in providing credits, foodstuffs, raw materials, and munitions to Britain, France, and Italy, it was able to ensure that the latter's military efforts *and* their production of armaments were maintained at high levels without financial embarrassment or bankruptcy. In other words, the United States now assumed much the same roles as Britain had occupied in the first two years of the war, providing the Allies with credit whilst steadily building up its own armed forces—all of which meant that the pressures upon the Central Powers would be kept up, and even increased, despite the defection of Russia.

Once again, the eve-of-war statistics help to illustrate the quite staggering imbalance which would occur if the German–Austrian *bloc* had to face an Anglo–American–French combination, even without Russia (see Table 10).

---

24. "Continental" Europe, because the British had already felt the impact of American expansionism from the 1890s onward.
25. Farrar, *Arrogance and Anxiety*, p. 39, fn. 17. The actual shift in the manufacturing production of Europe and the United States is detailed in Derek H. Aldcroft, *From Versailles to Wall Street: The International Economy in the 1920s* (Berkeley and Los Angeles: University of California Press, 1977), p. 98, table 4.

Table 10. Industrial/Technological Comparisons with the U.S. but without Russia

	U.K./U.S./France	Germany/Austria–Hungary
Percentage of world manufacturing production (1913)	51.7	19.2
Energy consumption (1913), metric million tons of coal equivalent	798.8	236.4
Steel production (1913), in million tons	44.1	20.2
Total industrial potential (1913), (U.K. in 1900 = 100)	472.6	178.4

In every economic category, the Anglo–American–French coalition was between two and three times as strong as Germany and Austria–Hungary combined—a fact confirmed by further statistics of the "war expenditures" of each side between 1914 and 1919: 60.4 billion dollars spent by the German–Austrian alliance, as opposed to 114 billion dollars spent by the British Empire, France, and the United States together (and 145 billion dollars if Italy and Russia's expenditures are included).[26] The advantages possessed by the Central Powers—of good internal communications, of the occupation (and exploitation) of enemy territories, of the defeat of Russia, of the fighting qualities of the German soldier—could not over the long run outweigh this massive disadvantage in sheer industrial muscle. And the average *Frontsoldat*'s amazement at how well provisioned were the Allied units which they overran during Ludendorff's last great offensive of spring 1918 was simply another confirmation of this disparity in economic forces.[27]

While it would be going too far to claim that the outcome of the First World War was completely predetermined, the above figures suggest that the overall course of that conflict—the early stalemate between the two sides, the ineffectiveness of the Italian entry, the slow exhaustion of Russia, the decisiveness of the American intervention in keeping up the Allied pressures, and the eventual collapse of the Central Powers—correlates closely to the economic and industrial forces available to each alliance during the different

26. Hardach, *The First World War, 1914–1918,* p. 153.
27. See the frequent anecdotes in Martin Middlebrook, *The Kaiser's Battle: 21 March 1918, The First Day of the German Spring Offensive* (London: Allen Lane, 1978).

phases of the struggle. To be sure, generals still had to direct (or *mis*direct) their campaigns, troops still had to summon the individual moral courage to assault an enemy position, and sailors still had to endure the rigors of sea warfare; but the record indicates that strategical insight and personal bravery were common to *each* side, and not enjoyed in disproportionate measure by one of the coalitions. What *was* enjoyed by one side was, particularly after 1917, a marked superiority in productive forces.

Economic Factors in the Origins and Effects of the War

In the light of this analysis of the long-term, secular trends in world politics from 1880 onwards, it may be useful to return to some of the questions which have exercised historians and other commentators ever since the First World War itself: for example, which power was the chief "challenger" (or "threat") to the status quo? What was the effect of the war on the great-power constellation? Where does the 1914–1918 conflict fit into the larger story of the *relative* decline of Europe?

The debate about the country which posed the chief challenge to the status quo before 1914 has always been a contentious one, since it carried with it connotations of guilt for causing the war itself and all its attendant losses.[28] The approach adopted in this essay is less concerned with political morality and national justifications than with the evidence one might reasonably draw from the indices of state power given above. The danger of such a line of argument is precisely that it may play down the political responsibility of the leaders of a country which was economically declining in relative terms (one thinks here of the Austrian "hardliners," or the French *revanchiste* lobby), which in a complete analysis of the causes of World War I it would be quite one-sided to do.[29] Fear of national eclipse, and of political decline, was often a more powerful motive for action than aggressive confidence about the future.

---

28. For this debate, made the more heated after Fritz Fischer's publications, see H.W. Koch, ed., *The Origins of the First World War* (London and New York: Macmillan, 1972); and John A. Moses, *The Politics of Illusion: The Fischer Controversy in German Historiography* (London: George Prior, 1975).
29. Especially since French apprehensions of further decline mingled with ambitions about recovering Alsace–Lorraine, just as Austrian fears about the collapse of the Dual Monarchy coexisted with desires to crush the South Slavs.

Since the United States' own growth was so widely ignored, only two countries were really perceived in the pre-1914 political debate as threats to the *general* equilibrium—perceived, in other words, as likely to enjoy further expansion of national power, leading sooner or later to territorial acquisitions in Europe or the Middle East, and to a radical change in the rough balance between the alliance blocs which had existed since the Franco–Russian treaties of the early 1890s. The first of these was Imperial Germany. Arguably it was not so much that country's booming population, industrial production, and overseas trade which concerned outside observers, even if those features were recognized as reflecting German *power;* it was, rather, the unpredictable actions of the Kaiser, the unscrupulous diplomacy of Bülow, the inexorable rise of the High Seas Fleet, and the threats used by Berlin during the Moroccan and Bosnian crises—that is, the "style" of German political actions—which alarmed its neighbors. And because there was no coherent decision-making process at the top, and because some German interests were attempting to penetrate the Balkans or Turkey or Luxemburg or Central Africa while others concentrated upon the naval race in the North Sea or the arms buildup on land, the impression was given of a country with immense ambitions and a Pan-Germanic "plan" for annexations in Europe and overseas.[30]

To some degree, these fears of German ambitions eased after 1911 (although not, of course, in the minds of the pathological anti-Germans in Entente countries). Much of the public excitement over the Anglo–German "naval race" had died away by then. Cooperation between Berlin and London during the Balkan crises suggested that the antagonism was not irrevocable and absolute. Bethmann Hollweg's evident dislike of Pan-Germanism, with what he called its "strident, pushing, elbowing, overbearing spirit," helped to assuage foreign suspicions. Above all, perhaps, the *military* balance of power seemed far less heavily tilted in Germany's favor than it had in 1905 (when, with Russia exhausted by the war with Japan, the German army could fairly easily have overrun France) or even in 1908–09 (when the Russian backdown over Bosnia–Herzogovina clearly signalled that its armed forces were still unprepared for major war).

30. I have discussed this point further in my *The Rise of the Anglo–German Antagonism,* especially pp. 430–431 and Conclusion.

By 1912, indeed, many observers were pointing to signs of Russian might rather than weakness, and the Prussian General Staff was becoming alarmed at the sheer dimensions of the military buildup in the East. The younger Moltke's fear that by 1916 or 1917 "our enemies' military power would then be so great that he did not know how he could deal with it" naturally caused him to urge a preventive war before it was too late;[31] but this reaction, however extreme, was merely one sign among many of the growing apprehension at Russia's military-industrial expansion.[32] It is in this sense that one can also understand the comment of Buchanan, the British ambassador to St. Petersburg, in April 1914 that "Russia is rapidly becoming so powerful that we must retain her friendship at almost any cost."[33] Although other Britons would have drawn the opposite conclusion to Buchanan's,[34] the fact remains that the Czarist Empire was regarded by many as a growing force in world affairs.

This perception presents historians with certain problems which have not (to my knowledge) adequately been explained in studies upon the origins of the war. The first puzzle is, why was Russian power before 1914 so absurdly overrated? Of course one can point (as in the paragraphs above) to the enormous *size* of the Russian army, but it had always possessed forces far larger than any of the other European states and that numerical preponderance had not prevented its dismal showing in the Crimean War, the Russo–Turkish War of 1877, and the Russo–Japanese War. Why was it expected to do much better in 1914, and against an enemy as formidable as Imperial Germany? Had the strategic "experts" of the time become mesmerized by *quantity*, rather than quality? Did they not see that Russia, despite its lurch towards industrialization, was a military colossus but an economic pygmy?

---

31. Fischer, *War of Illusions*, p. 402.
32. See Ropponen, *Die Kraft Russlands*.
33. Cited by K. Wilson, "British Power in the European Balance, 1906–1914," in David Dilks, ed., *Retreat from Power: Studies in Britain's Foreign Policy of the Twentieth Century*, 2 vols. (London: Macmillan, 1981), Vol. 1, p. 39.
34. The problem with Wilson's thesis (ibid.), that Britain really went to war in 1914 in order to protect India from future Russian pressures, lies not in his citations; clearly, there were a few, like Buchanan and Nicolson, who were fearful of Russia. But many more of the decision-makers (Henry Wilson, Crowe, Churchill, Haldane, and arguably Grey) were much more impressed by the German threat closer to home. In any case, the argument that Britain had to join the European war in order to please Russia (or out of fear of Russia) would have played into the hands of the anti-interventionists, who were virtually all hostile to Russia on ideological/political grounds.

Did those who pointed in alarm to Czarist expansionism not catch any sense of the apprehensions felt by the Russian leadership at its many weaknesses?

A second, related problem is this: which power, Russia or Germany, was *objectively* the most likely to alter the existing order in Europe? (*Subjective* fears, although important in their own right, are not at issue here, and if studied in detail produce a confusing picture: some Russians disliked the Anglo–Saxon powers more than they did Germany; some Austrians feared their German partner's ambitions almost as much as they did Russia's; some Britons feared Russia more than they did Germany, some Germans Britain more than Russia.) It has become a commonplace to assert that "the prodigious development of Russian armaments threatened to alter the whole balance of European power";[35] and even those historians who argue that Germany was making a "grasp for world power" readily point to German alarm at Russian military increases as constituting a major factor in Berlin's decision for war in 1914.[36] All this overlooks the staggeringly important fact that, despite all of Russia's absolute increases, its productive strength was actually *decreasing* relative to Germany's. Between 1900 and 1913, for example, its steel production rose from 2.2 to 4.8 million tons; but Germany's leapt forward from 6.3 to 17.6 million tons. In the same way, the increases in Russia's energy consumption and total industrial potential were not as large either absolutely *or* relatively, as Germany's. Finally, it will be noticed that in these years Russia's share of world manufacturing production *sank,* from 8.8 percent to 8.2 percent, due to the expansion of the German and (especially) the American shares.

That some Germans worried about the Russian "threat" is undoubted. But what is also undoubted, as A.J.P. Taylor acutely pointed out some 30 years ago, was that:

> Their fears were exaggerated. Certainly, Russia would have been a more formidable Power by 1917, if her military plans had been carried through and if she had escaped internal disturbance—two formidable hypotheses. . . . in any case, the Russians might well have used their strength against Great Britain in Asia rather than to attack Germany, if they had been left alone. In fact, peace must have brought Germany the mastery of Europe within a few years.[37]

---

35. Turner, *Origins of the First World War,* p. 75. For a more sober assessment of Russian power at that time, see Lieven, *Russia and the Origins of the First World War.*
36. See Fischer's *War of Illusions,* p. 370 ff.
37. Taylor, *The Struggle for Mastery in Europe,* p. 528.

And even if that scenario of an Anglo–Russian war in Asia did not occur, the figures suggest that Russia would do well to maintain its position *vis-à-vis* Germany, let alone improve it.

One final puzzle remains. Since the antagonism between the various powers seemed so deep-rooted, and was portrayed in the Social Darwinistic language of the time as both fundamental and inevitable, it remains curious that each combatant believed that the war, when it came, would be short. Those, like Bloch, who suggested otherwise were disregarded or openly attacked. Exactly why men so passionately believed in a swift, decisive victory has received many explanations, each of which helps in a way to provide an answer. Nevertheless, to this day the pre-1914 age presents that strange dichotomy between, on the one hand, the widely held perceptions of fundamental, almost cosmic trends which seemed to face each of the powers with the choice of *Weltmacht oder Niedergang;* and, on the other, the bland assumption that the conflict would soon be over. Yet few if any commentators appear to have found it odd that what was forecast as an epic "struggle for survival" was generally expected to be decided in one season's campaigning.

Effects of the War on International Politics

For many years, it has been assumed that the First World War quite altered the great-power constellation. In one significant way, that is, in the collapse of Austria–Hungary and its replacement by various "successor-states," the war's impact was decisive. But with the passing of time, and the coming of a longer historical perspective which sets the events of 1914–1918 within global trends of the past 100 or 150 years, the innovatory and cataclysmic nature of the Great War seems less pronounced, at least in terms of international politics.[38]

It is true that the war, in helping to cause the collapse of authority in Russia (and thereby giving a ruthless pragmatist like Lenin an opportunity to short-cut the Marxist path towards socialist revolution), produced an alteration in global politics which has overshadowed the rest of the twentieth century. Yet granted the magnitude of the *ideological* impact of the October

38. Psychologically, of course, its impact was immense, and in countries like Britain and France lingers to this day. But psychological shock often leads to *mis*perception of political and military realities.

Revolution upon the world, one is still tempted to ask the following questions:

1. Was it likely, had there not been a First World War, that Russia's strained sociopolitical order would have avoided internal convulsions as the pace of industrialization became more rapid? And would any new *regime,* aware of the fissiparous movements that existed from Finland to the Ukraine, have gracefully surrendered the "Great Russian" traditions of internal monopolization and external aggrandizement?

2. Even had the old order—autocratic, expansionist, occasionally messianic, and always nationalistic—remained in power, and had it avoided a First World War, what further impact would Russia have made upon the world as its vast resources were steadily modernized? Viewed against the long-term "trajectory" of Russian power in the world, it is clear that the 1914 conflict came at the worst possible time. In the eighteenth century, due to the standardization of weapon types and armies, Russia had been able to pull itself up to great-power status by borrowing heavily from the West and using its sheer advantage in numbers; in 1815 it already was *the* great military power in Europe. However, for most of the nineteenth century it had lost ground, *relatively,* because industrialization gave incredible advantages of productivity and wealth to countries like Britain, and later Germany. What Russia required after 1900 was a number of decades of peaceful modernization in order to close the gap between its actual backwardness and its great *potential* power—and, by extension, to catch up on the West once again. This drive towards modernization was obviously distorted by the war, and it would be perverse to deny the latter's impact. Nevertheless, the Bolshevik successor regime was as anxious to speed up that process (and with the same heavy-handed preference for large industrial plants and armaments-related goods) as the Czarist Council of Ministers before 1914.

In other words, regardless of the enormous political and ideological consequences of the Bolsheviks' success, one suspects that the late twentieth-century world would have felt the impact of Russian might in any case.

But did not the First World War at least deal a fatal blow to France and Britain's place in the world? That the losses in manpower, wealth, and national morale in both countries were prodigious is very clear. Yet the fact is that, in France's case, the country was being very swiftly eclipsed as an independent Great Power simply by Germany's peacetime expansion during the decades before 1914. In 1880, France was just a little inferior to its eastern neighbor in its share of world manufacturing production, its iron/steel output, its energy consumption, and its total industrial production; by the eve of war, Germany was between *two and three times* as powerful. What the war did, albeit at great cost to the French themselves, was to check the German bid for mastery in Europe and to restore France to its pre-1870 position as the largest military nation in Western Europe. That position was, of course, an artificial one and as soon as the innate strength of the German economy began to recover from a decade-and-a-half of war, revolution, inflation, and slump, France was once again eclipsed. By 1938 its position of economic and industrial inferiority relative to Germany was perhaps slightly worse than in 1913; but could one plausibly argue that the relative position would have been better had the First World War not occurred? It seems unlikely.

Is it, then, too absurd to make roughly the same argument in Britain's case? The strains of war—upon its global financial position, upon its domestic politics, upon intra-imperial relations, upon Ireland and India—are well known.[39] In consequence, perhaps, historians forget how much more favorable to British interests the power constellation was in 1914 than, say, in 1900, when imperialists like Milner had wondered whether it was possible to "save the Empire" and members of the Prussian General Staff had talked of the coming "War of British Succession."[40] It was true that the alarmingly swift rise of Imperial Germany challenged British interests in economic and colonial terms, in its naval security, and ultimately in its concern for a balance of power in Europe; but a clumsy Wilhelminian *Weltpolitik* also managed to provoke France, Russia, and even (distantly[41]) the United States. As the statistics above (pp. 21 and 25) show, Britain fought in the First World War as part of an overwhelming coalition. Would it have fared better had there occurred merely an Anglo–German war? or a war between Britain and the Dual Alliance? or an Anglo–American war? or a war where Britain was alone against a *combination* of other powers? In the diplomatic circumstances around 1900, none of those scenarios seemed impossible, and some seemed very likely.[42] In sum, any argument that the war weakened Britain's world position

39. See the brief summary in Paul M. Kennedy, *The Realities Behind Diplomacy: Background Influences on British External Policy, 1865–1980* (London and Boston: Allen & Unwin, 1981), part II.
40. Kennedy, *The Rise of the Anglo–German Antagonism*, chapters 13, 16.
41. See Herwig, *Politics of Frustration*, passim.
42. See footnote 40 above; and George W. Monger, *The End of Isolation: British Foreign Policy, 1900–1907* (London: Thomas Nelson, 1963), chapter 1.

needs to grapple with the counter-proposition that, *objectively*, its position had been shored up by the defeat of Germany, the Bolshevik revolution, and the end of the Turkish empire. As the German historian Erich Marcks noted bitterly in the winter of 1920/21,

In the world Russia and Germany have now collapsed, a colossal gain for England. . . . She has secured the double aim of her imperialism, to dominate the route from Cairo to the Cape, and from Cairo to Calcutta . . . the Indian ocean in its totality has become an English sea. In India itself England has deployed powerful forces. She has strengthened her power and her trade, has gained valuable new regions in Mesopotamia, Persia and Africa, and her world empire has increased in land-size by around 27 per cent and in population by almost the same. This has resulted in a global power and position as never before; England is the only winner from this war, England together with North America: one can see an Anglo–Saxon world mastery rising on the horizon. . . .[43]

As many authors (including the present one) have pointed out, this apparently strong position was also artificial, like France's. It depended upon the (temporary) suppression of German power in Europe, upon the (temporary) elimination of Russia as a force in world affairs, upon (temporary) Japanese abstention from further expansion in the Far East, and—as much as anything—upon the United States' remarkable return to quasi-isolationism after 1920 despite its unequaled economic preeminence. By the late 1930s, at the latest by the early 1940s, all of those latent forces for global change had returned, with a greater impact than before; and the end of the British Empire could no longer be postponed. But all this does not obviate the fact that the 1914–1918 struggle relatively enhanced Britain's international position, allowing it to enjoy the fruits of its eighteenth and nineteenth-century expansionism for a few decades more.

For Germany, by contrast, the war was a setback in its drive to become one of the really great world powers. The combination of states opposed to it was, quite simply, too much. Yet its overall performance, as Northedge points out, was extraordinary:

for the four and half years of the First World War, Germany, with no considerable assistance from her allies, had held the rest of the world at bay, had beaten Russia, had driven France, the military colossus of Europe for

43. Cited in Kennedy, *The Realities Behind Diplomacy*, p. 223.

more than two centuries, to the end of her tether, and, in 1917, had come within an ace of starving Britain into surrender.[44]

It was one of the ironies of the peculiarly assertive nature of German *Grossmachtpolitik* that the defeat of 1918, far from inducing cautious strategies suitable for any later "war of recovery," helped the rise of Hitler and the adoption of even more extreme foreign and military policies—provoking once again the creation of an overwhelming alliance to bring Germany down. Yet while the unconditional surrender of 1945 was altogether more decisive than the previous defeat, it is worth noting that within a further three decades the truncated "West" Germany was again among the leading three or four economic giants in the world; and, did it occupy its 1937 borders, the united German nation would today be a larger industrial and economic force than the Soviet Union. In the grand sweep of military and economic developments between 1880 and 1980, the 1914–1918 conflict was the first, almost (but perhaps not quite) irrevocable blow dealt to what otherwise might have been the steady German acquisition of primacy in Europe. What Hitler's manic ambitions did was to eliminate the chances of a real recovery, and to move the victorious Allies to "solve" the "German question" by partly undoing the 1871 settlement.

The case of the United States provides a further, final example to sustain the argument that the Great War may have had less impact upon the long-term power balances than has commonly been supposed. Of course, the conflict led to the mobilization of American resources for military purposes in a way that peacetime conditions would never permit (just as the 1941–1945 war led to an even greater harnessing of the United States' productive forces). Of course, the American support for the western Allies, before as well as after 1917, was crucial. Of course, the hostilities in Europe led to a great acceleration in U.S. warship building, so that it would swiftly have a fleet "second to none," a full rival to the Royal Navy. And, of course, the war led to a shift in the economic center of gravity, from one side of the Atlantic to the other. But the question to be faced is, would this not have occurred *in any case*? Whenever it wanted after 1900, the U.S. could probably have outbuilt the Royal Navy. And, as noted above, it was becoming the

44. F.S. Northedge, *The Troubled Giant: Britain among the Great Powers, 1916–1939* (London: Bell/London School of Economics, 1966), p. 623.

industrial center of the world even before 1914; while detailed studies of the international gold, credit, and investment systems suggest that the financial center of gravity was also shifting across the Atlantic.[45]

It may be rash to go further than this; but it seems to this writer that the technological and other factors leading to the United States' economic expansion after 1800 were *so* favorable—rich agricultural land, vast raw materials, no social or geographical restraints, the evolution of technology (the railway, the steam engine) to exploit such resources, the absence of foreign dangers—that there was a virtual inevitability about the process: that is to say, only persistent human ineptitude, or near-constant civil war, or a climatic disaster could have checked this rise to global economic (and, by extension, military) influence. Precisely what combination of factors causes nations to rise and fall, and economies to expand and contract, is not for analysis here. But that they do rise and, despite the frantic efforts of political leaders, also relatively decline later is incontrovertible: new centers of production, often exploiting newer techniques, do over time cause substantial shifts in the economic balances. It ought to be of some interest to today's readership, therefore, to reproduce the "trajectory" of the United States' place in the global economy[46] (see Figure 1). What this graph suggests, *inter alia*, is the relatively small impact of the First World War upon America's changing global position. Boosted by its own agricultural and industrial revolutions, it had already surged forward to overtake Britain as the greatest economic force in the world. The 1914–1918 period may have accelerated that drive; just as the international slump of the 1930s reduced it (since the drop in purchasing power, and the widespread protectionism, made it more difficult than before for American products to overwhelm foreign markets). Later, because the Second World War was so very devastating—not merely for the German and Japanese economies, but also for "victors" such as Britain, the Soviet Union, and France—the U.S. share of world manufacturing production rose to a peak. Since then it has steadily declined, partly because that peak was somewhat artificial, partly because new centers of production

---

45. Marcello De Cecco, *Money and Empire: The International Gold Standard, 1890–1914* (Oxford: Oxford University Press, 1974); Benjamin M. Rowland, ed., *Balance of Power or Hegemony: The Interwar Monetary System* (New York: New York University Press, 1976).
46. Figures up to 1980 come from Paul Bairoch, "International Industrialization Levels from 1750 to 1980," *Journal of European Economic History*, Vol. 11, No. 2 (Spring 1982), passim. The projection to 2000 is one given in a number of studies, and used from time to time in *The Economist*.

Figure 1. U.S. Manufacturing Production as a Percentage of World Production

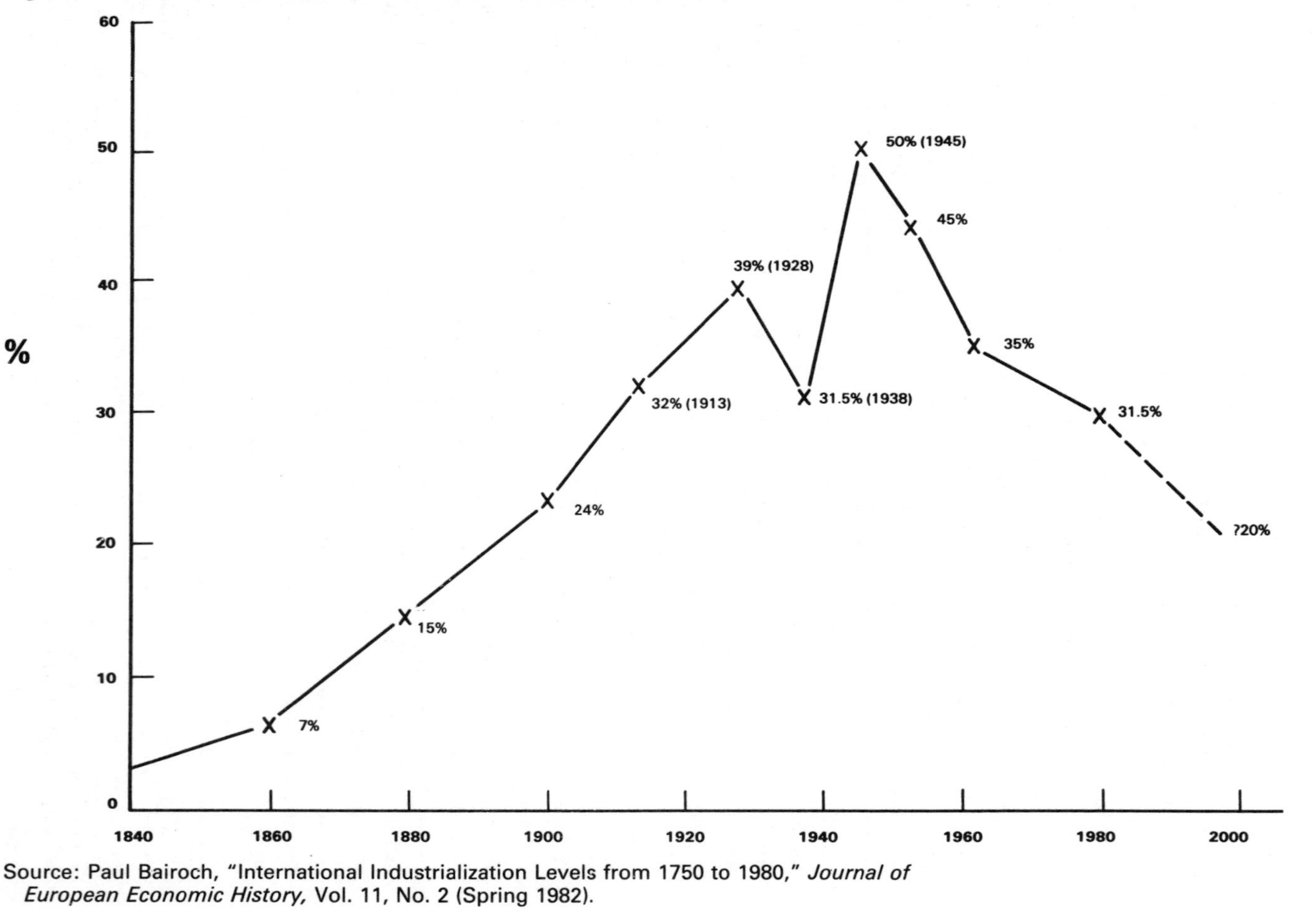

Source: Paul Bairoch, "International Industrialization Levels from 1750 to 1980," *Journal of European Economic History*, Vol. 11, No. 2 (Spring 1982).

have arisen in the Far East and elsewhere. All this may be taking us a long way from the First World War; but there is nothing like the *longue duree* to remind historians to be careful of claiming that any particular event "revolutionized" the affairs of men.

Concluding Remarks

To pinpoint "lessons" for the present-day nuclear world from what has been argued above seems a rash enterprise. As with most complex happenings, the First World War offers so much data that conclusions can be drawn from it to suit any *a priori* hypothesis which contemporary strategists and politicians wish to advance. Since it is not clear whether the existence of nuclear weapons has totally revolutionized the nature of great-power conflict—and that question can only be "answered" in the affirmative by the use of such weapons, leading perhaps to everyone's destruction—it becomes still more of an uncertain exercise to study history as a guide to present-day policy. The few remarks which follow are therefore to be regarded as tentative, suggestive, provisional—and *not* in any way to be seen as firm conclusions.

The first thing to wonder about is the recurring belief held by each generation that the next major war, whenever it came, would be a *short* one. There had been, of course, certain conflicts in the late-nineteenth century which were swiftly over: the Bismarckian wars against Denmark, Austria, and France; the Sino–Japanese war; and the Spanish–American war. But they had all involved a clash between two powers alone, in which one of them possessed considerable military advantages and was not hindered by third powers. When each side was more evenly balanced, as in the early stages of the American Civil War, or the Russo–Japanese war, the prospects of a swift victory were less, especially if the terrain was extensive; and when a conflict occurred between two large coalitions, as in 1914, it was almost bound to be a lengthy, grinding affair—just as the coalition wars against Napoleon had been.

As we know, all of this was ignored by "experts" before 1914 who held that new technology and organization—artillery, railways, mass armies, all-big-gun battleships—had given the clear advantage to the offensive, and that even major wars would be over within months. The dreadful mutual hemorrhaging of each side's youth in the stalemated war of 1914–1918 may have checked that assumption for a while, but the advent of the new "wonder weapon," the bomber, caused a revival in the belief that a future war would be swiftly concluded. Yet the Second World War—another grand coalition war—was to last six years before the latent economic/industrial strength of one of the *blocs* was fully converted into overwhelming military advantage. Thus, while conceding that nuclear weapons may indeed have transformed the nature of warfare, it nonetheless seems worthwhile to wonder if the prevailing assumption about the length of any future *conventional* struggle between the forces of NATO and those of the Warsaw Pact—that is, the assumption that everything will be decided swiftly, in an epic clash of tanks and aircraft over the north German plains—is not a repetition of older mistakes? Has any military planning group or research institution tried to estimate the possible implications of a *lengthy* major war, and of the demands which would be made upon the combatants after (say) two years of hostilities? If that seems an utterly stupid question to ask, then it is worth recalling that it must have seemed equally stupid in 1914 to all the general staffs and politicians concerned.

The second and related general point concerns the economic underpinnings of war. Since that is a theme of great interest to this writer,[47] its importance has been emphasized in the above analysis—to the detriment, no doubt, of such factors as leadership, national morale, etc. Yet it is not an exaggeration to claim that major, long-lasting wars between great-power coalitions have *always* been won, in the modern age, by the side with the largest economic staying power and productive resources. *Pas d'argent, pas de Suisses* may have been a motto of Renaissance diplomacy, but the point remains valid to this day. Without a flourishing and efficient economic base, and without the capacity to keep on raising funds, a nation's military power lacks long-term credibility.

By extension, then, the politicians and strategists of the leading countries today might do well to concentrate rather more upon the often subtle relationships between military strength and economic strength—and to wonder whether, in some instances, front-line military and naval forces are the best measure of a nation's real power. To be sure, all this once again implies that future conflicts may be long, not short; but if that assumption turns out to be true, and if industrial productivity, technological efficiency, and economic staying power are going to be as important in the future to a nation's place

47. This essay, although written at invitation to fit into the present issue of *International Security*, overlaps to some extent with a book I am presently writing on the relationship between the economic and military strengths of the great powers, 1500 to 2000.

in world affairs as they have been in the past, then two further hypotheses can be advanced: one of comfort to the West, the second less so. The first is that, unless and until the Soviet Union makes significant (and unlikely?) rapid advances in scientific technology and entrepreneurial efficiency, it is going to remain militarily top-heavy for its economic base—a nice irony, given the Marxist conception of history. It is true that this top-heaviness is probably not as acute as in 1914, and a "closed" society will always find it easier to allocate more upon guns than upon butter; but various indicators (from the excessive share of GNP devoted to armaments, to the backwardness in computer technology[48]) suggest a dangerous imbalance on the part of the Soviet Union should a conventional war occur and last for several years at least.

The second hypothesis is that, notwithstanding the quite staggering increases in the monies allocated to the U.S. armed forces under the Reagan Administration, which intends to "reassert" the country's place in world affairs, the long-term position is less than rosy. Once again, it will not do to overstate the case, or to turn this into more than a hypothesis. But a nation possessing 20 percent of the world's manufacturing production (as the United States may do in the year 2000) obviously has far less economic/industrial muscle relative to others than when it held 35 percent or 50 percent of the whole. It may be some slight consolation to American "hawks" that the redistribution of the global economic balances indicates no real shift to the Soviet Union, which is also losing ground relative to Japan and certain Third World countries. But it cannot conduce to the long-term health of the American economy to be falling behind in some of the vital battles in advanced technology;[49] to have its present enormous budgetary and visible-trade deficits being covered by the inflow of volatile foreign-held investments, which may just as swiftly move out of the dollar unless interest rates remain high; and to learn that the damage being done to American export industries by the consequent overvaluation of the dollar may be difficult (or impossible) to repair even if the currency returned to a more realistic level in the future.

All this is not to deny the very real strengths which the United States possesses, both in economic and military terms; but it is *less* strong, relatively,

---

48. Marshall I. Goldman, *U.S.S.R. in Crisis: The Failure of an Economic System* (New York: W.W. Norton, 1983).

49. As I write this in draft, *The New York Times* (February 13, 1984) reports on "Big Japanese Gain in Computers Seen: Scientists Fear New Devices will end U.S. Dominance," a not uncommon occurrence.

than thirty years ago and at the same time its overseas commitments have increased alarmingly. In such a situation, perhaps a little less attention ought to be given to, say, B-1 bombers; and instead, a little more notice should be paid to the way the leaderships of previous world powers grappled with foreign and defense policy problems when they, too, were losing their relative economic primacy and realizing the need to arrange more carefully the complex and intimate links between military strategy, armaments production, and the nation's industrial and financial base.[50] In this respect, the Pentagon might do well to heed an earlier expert on military affairs, who many years ago warned: "Nothing is more dependent on economic preconditions than precisely the army and the navy."[51]

---

50. They could begin with David French, *British Economic and Strategic Planning, 1905–1915* (London and Boston: Allen & Unwin, 1982). See also "Strategy *versus* Finance in Twentieth-Century Britain," in Kennedy, *Strategy and Diplomacy*, pp. 87–106.
51. Friedrich Engels, *Herr Eugen Duhring's Revolution in Science* (London, 1936), p. 188.

Men Against Fire

Expectations of War in 1914

Michael Howard

In 1898 there was published in Paris a six-volume work entitled *La Guerre Future; aux points de vue technique, economique et politique.* This was a translation of a series of articles which had been appearing in Russia, the fruit of collective research but masterminded and written by one of the leading figures in the world of Russian finance and industry, Ivan (or Jean de) Bloch (1836–1902). Sometimes described as "a Polish banker," Bloch was in fact an entrepreneur almost on the scale of the Rothschilds in Western Europe or Carnegie in the United States. He had made his money in railroad promotion, and then turned to investment on a large scale, promoting and sharing in the great boom in the Russian economy of the 1890s. He had written prolifically about the economic problems of the Russian Empire, and was increasingly alarmed by the degree to which they were complicated, then as now, by the military need to keep abreast, in an age of rapidly developing technology, with the wealthier and more advanced states of the West. Having been responsible for organizing the railway supply for the Russian armies in their war with the Ottoman Empire in 1877–78, Bloch had an unusual grasp of military logistics. And he brought to the study of war an entirely new sort of mind, one in which the analytical skills of the engineer, the economist, and the sociologist were all combined. His book was in fact the first work of modern operational analysis, and nothing written since has equalled it for its combination of rigor and scope.

Only the last of the six volumes was translated into English, under the title *Is War Now Impossible?*[1] This volume conveniently summarizes the argument of the entire work, and it was itself summarized by the author in an interview with the English journalist W.T. Stead which is printed as an introduction to the book. Bloch began by stating his conclusions: war between great states was now impossible—or, rather, suicidal. "The dimensions of modern armaments and the organisation of society have rendered its pros-

Michael Howard is Regius Professor of History at Oxford University. This paper was written while he was a Fellow at the Woodrow Wilson International Center for Scholars in Washington, D.C.

1. Jean de Bloch, *Is War Now Impossible? The Future of War in its Technical, Economic and Political Relations* (London and Boston, 1899).

International Security, Summer 1984 (Vol. 9, No. 1) 0162-2889/84/010041-17 $02.50/1

ecution an economic impossibility."[2] This could be almost mathematically demonstrated. The range, accuracy, and rate of fire of modern firearms—rifles lethal at 2000 meters, artillery at 6000—made the "decisive battles" which had hitherto determined the outcome of wars now impossible. Neither the infantry could charge with the bayonet nor cavalry with the saber. To protect themselves against the lethal storm of fire which would be unleashed on the modern battlefield, armies would have to dig themselves in: "the spade will be as indispensable to the soldier as his rifle. . . . That is one reason why it will be impossible for the battle of the future to be fought out rapidly. . . . Battles will last for days, and at the end it is very doubtful whether any decisive victory can be gained."[3]

Thus far Bloch was not breaking new ground. He was only setting out a problem which intelligent officers in all European armies had been studying ever since the experiences of the Franco–Prussian War in 1870 and the Russo–Turkish War in 1877–78 had shown (quite as clearly as, and rather more immediately than, those of the American Civil War) the effect of modern firearms on the battlefield. The introduction of "smokeless powder" in the 1880s, increasing the range and accuracy of all firearms and making possible the near invisibility of their users, would, it was generally agreed, complicate the difficulties of the attack yet further. But even these, it was widely assumed, would not change the fundamental nature of the problem.

The answer, it was believed, lay in the development of the firepower of the assailant, especially of his artillery. The assaulting infantry had to approach closely enough, making all use of cover, to be able to deploy a hail of rifle fire on the defenders' positions. Artillery must cooperate closely, keeping the defenders' heads down with shrapnel and digging them out of their trenches with high explosives. As for machine-guns, these, with their mobility and concentrated firepower, were seen as likely to enhance the power of the attack rather than the defense. "Fire is the supreme argument," declared Colonel Ferdinand Foch in his lectures at the École de Guerre in 1900.[4] "The superiority of fire . . . becomes the most important element of an infantry's fighting value." But the moment would always come when the advance could get no further: "Before it is a zone almost impassable; there remain no covered approaches; a hail of lead beats the ground . . . to flee or

---

2. Ibid., p. xi.
3. Ibid., p. xxviii.
4. Ferdinand Foch, *The Principles of War* (New York, 1918), p. 362.

to charge is all that remains." Foch, and the majority of French thinkers of his time, believed that the charge was still possible and could succeed by sheer dint of numbers: "To charge, but to charge in numbers, therein lies safety. . . . With more guns we can reduce his to silence, and the same is true of rifles and bayonets, if we know how to make use of them all."[5] Others were less sure. The Germans, who still after thirty years had vivid memories of the slaughter of their infantry at Gravelotte, preferred if possible to pin the enemy down by fire from the front but attack from a flank. Nobody was under any illusion, even in 1900, that frontal attack would be anything but very difficult and that success could be purchased with anything short of very heavy casualties. There would probably indeed have been a wide measure of agreement with Bloch's calculation, that a superiority at the assaulting point of 8 to 1 would be necessary to ensure success.[6]

Bloch's War of the Future: Society versus Society

It was in the further conclusions which Bloch deduced from his study of the modern battlefield that he outpaced his contemporaries—not so much because they disagreed with him, but because they had given the problems which he examined virtually no thought at all.

What, asked Bloch, would be the eventual result of the operational deadlock that was likely to develop on the battlefield? "At first there will be increased slaughter—increased slaughter on so terrible a scale as to render it impossible to push the battle to a decisive issue. . . . Then, instead of a war fought out to the bitter end in a series of decisive battles, we shall have to substitute a long period of continually increasing strain upon the resources of the combattants." This, would involve "entire dislocation of all industry and severing of all the sources of supply by which alone the community is enabled to bear the crushing burden. . . . That is the future of war—not fighting, but famine, not the slaying of men but the bankruptcy of nations and the break-up of the whole social organisation."[7] In these circumstances the decisive factors would be "the quality of toughness and capacity for endurance, of patience under privation, of stubborness under reverse or disappointment. That element in the civil population will be, more than

5. Ibid., pp. 365–366.
6. Bloch, *Is War Now Impossible?*, p. xxvii.
7. Ibid., p. xvii.

anything else, the deciding factor in modern war. . . . Your soldiers," concluded Bloch grimly, "may fight as they please; the ultimate decision is in the hands of *famine*."[8] And famine would strike first at those proletarian elements which, in advanced industrial societies, were most prone to revolution.

It is important to recognize that Bloch got a great deal wrong. He assumed that the prolonged feeding and administration of the vast armies which rail transport made possible would be far beyond the capacity of the military authorities, and that armies in the field would quickly degenerate into starving and mutinous mobs. He predicted that the care of the sick and wounded would also assume unmanageable proportions, and that on the battlefield the dead and dying would have to be heaped up into macabre barriers to protect the living from enemy fire. As did many professional soldiers, Bloch doubted the capacity of reservists fresh from civil life to stand up to the strain of the battlefield: "it is impossible to rely upon modern armies submitting to sacrifice and deprivation to such an extent as is desired by military theorists who lose sight of the tendencies which obtain in Western society."[9] In fact the efficiency with which armies numbering millions were to be maintained in the field, the success with which the medical services were, with certain grisly exceptions, to rise to the enormous task that confronted them and the stoical endurance displayed by the troops of all belligerent powers in face of hardships worse than Bloch could ever have conceived were perhaps the most remarkable and admirable aspects of the First World War. Bloch, like so many pessimistic prophets (including those of air power a generation later), underestimated the capacity of human societies to adjust themselves to adverse circumstances.

But Bloch also had astonishing insights. The scale of military losses, he pointed out, would depend on the skill of the commanders, and "it must not be forgotten that a considerable number of the higher officers in modern armies have never been under fire"; while among junior officers the rate of casualties would, if they did their job as leaders, be inordinately high. Finally, there was the problem of managing the wartime economy; what were the long-term effects of that likely to be? "If we suppose," Bloch surmised, "that governments will be forced to interfere in the regulation of prices and to support the population, will it be easy after the war to abandon this practise

---

8. Ibid., p. xlvi.
9. Ibid., p. 30.

and re-establish the old order?"[10] Win or lose, therefore, if war came "the old order" was doomed—by transformation from above if not by revolution from below.

This remarkably accurate blueprint for the war which was to break out in Europe in 1914, last for four and a half years, and end only with the social disintegration of the defeated belligerents and the economic exhaustion of all was the result, not of second-sight, but of meticulous analysis of weapons capabilities, of military organization and doctrine, and of financial and economic data—five fat volumes which still provide a superb source book for any student of the military, technological, and economic condition of Europe at the end of the nineteenth century. Nobody took Bloch's economic arguments and attempted to disprove them. They were just ignored. Why, it may be asked, was so little account taken of them by statesmen and military leaders? Why did they continue on a course which led ineluctably to the destruction of the old order which Bloch so unerringly predicted? The question is one uncomfortably relevant to our own times.

The answer is of course that societies, and the pattern of international relationships, cannot be transformed overnight on the basis of a single prophetic insight, however persuasively it may be argued. Bloch's thinking and influence were indeed two elements in persuading Czar Nicholas II to convoke the first International Peace Conference which met at the Hague in May 1899, and were even more significant in mobilizing public support throughout Europe for that conference's objectives. But the conference was no more than a ripple in the current of international politics. A more immediate problem, as Bloch himself repeatedly pointed out, was that there existed nowhere in Europe bodies charged with the task of thinking about the problems of warfare in any kind of comprehensive fashion, rather than about the narrowly professional questions that concerned the military. As for the military specialists, they were not likely to admit that the problems which faced them were insoluble, and that they would be incapable in the future of conducting wars so effectively and decisively as they had in the past.

Lessons of the Boer War

The force of Bloch's arguments, however, was powerfully driven home when, within a few months of the publication of *La Guerre Future*, there broke

---

10. Ibid., pp. 335, 314.

out in South Africa a war in which for the first time both sides were fully equipped with the new technology—magazine-loading small-bore rifles, quick-firing artillery, machine guns—and things turned out on the battlefield exactly as he had predicted. The British army, moving in close formations and firing by volleys, were unable to get anywhere near an enemy whom they could not even see. At Spion Kop, at Colenso, at the Modder Rover, and at Magersfontein, their frontal attacks were driven back by the Boers with horrifying losses. As the leading British military theorist, Colonel G.F.R. Henderson, who accompanied the army in South Africa, wrote shortly afterwards:

> There was a constant endeavor to make battle conform to the parade ground . . . to depend for success on courage and subordination and to relegate intelligence and individuality to the background . . . the fallacy that a thick firing line in open country can protect itself, outside decisive range, by its own fire, had not yet been exposed. It was not yet realised that the defender, occupying ingeniously constructed trenches and using smokeless powder, is practically invulnerable to both gun and rifle.[11]

Unsympathetic continental observers tended to play down the significance of the South African experience on the grounds that the British army and its commanders were unsuitably trained for confronting a "civilized" adversary, having been spoiled by the easy victories in Egypt and the Sudan. Further, they suggested that the differences in terrain made the lessons to be learned from that war, as they had made those from the American Civil War, irrelevant in the European theater. The British themselves, while unable to deny the unsuitability of their traditional tactics and training to the transformed conditions of warfare, could nonetheless point out that, once they had mastered the necessary techniques, they had been able successfully to go over to the offensive, and had then rapidly won the war. This they had done by pinning down the Boers in their positions by firepower and maneuvering round their flanks with cavalry—cavalry used not in its traditional role for shock on the battlefield, but to develop the kind of strategic mobility which was essential if the problems created by the new power of the defensive

11. George F.R. Henderson, *The Science of War* (London, 1905), p. 411. It is ironic to read in an article which Henderson had written shortly before the war: "Neither smokeless powder nor the magazine rifle will necessitate any radical change. If the defense has gained, as has been asserted, by these inventions, the plunging fire of rifled howitzers will add a more than proportional strength to the attack. And if the magazine rifle has introduced a new and formidable element into battle, the moral element still remains the same." Ibid., pp. 159–160.

were to be overcome. When in 1901 Bloch described to an audience at the British Royal United Services Institution how the experience of the British army in South Africa, repeated as it would be in Europe on an enormous scale, precisely illustrated his arguments, his audience was able to point out that in fact Lord Roberts had shown how to combine the tactical advantages of firepower with the strategic advantages of horse-borne mobility to secure precisely those decisive results which Bloch had maintained would, in future, be impossible.[12]

A study of the voluminous military literature of the period shows that between 1900 and 1905 a consensus developed among European strategic thinkers over two points. The first was the strategic importance of cavalry as mobile firepower. If the firepower of the defense made it now impossible for cavalry to assault unshaken infantry—a view which had been reluctantly accepted ever since the disasters of the Franco–Prussian War of 1870—cavalry would now develop their own firepower, enhanced by mobile quick-firing artillery and machine guns, and exploit opportunities on a scale undreamed of since the days of the American Civil War. The South African experience indeed sent back intelligent cavalrymen, especially in England, to studying the Civil War, often for the first time.[13] In the British army, it was laid down that the carbine or rifle would henceforth be "the principal weapon" for cavalry. But for most cavalrymen this was going altogether too far. In no country in Europe was this proudest, most exclusive, most anachronistic of arms prepared to be, as they saw it, downgraded to the role of mounted infantry. That kind of thing could be left to colonial roughriders. Writing as late as 1912, the German general Friedrich von Bernhardi bitterly observed that "The cavalry looks now . . . upon a charge in battle as its paramount duty; it has almost deliberately closed its eyes against the far-reaching changes in warfare. By this it has *itself* barred the way that leads to greater successes."[14] Within the cavalry in every European army therefore a controversy raged which was settled only by the kind of compromise expressed by the British Cavalry Manual of 1907:

---

12. *Journal of the United Services Institution,* Vol. 15, pp. 1316–1344, 1413–1451.
13. G.F.R. Henderson had been writing and lecturing on the American Civil War well before 1899 and Lord Roberts was to acknowledge the influence of those writings on his own operational planning in South Africa. After 1901 the Civil War became the main topic for historical study at the British Army Staff College at Camberley. Jay Luvaas, *The Military Legacy of the Civil War* (Chicago: University of Chicago Press, 1959), p. 229.
14. Friedrich von Bernhardi, *On War Today* (London, 1912), Vol. 1, p. 192.

The essence of the cavalry spirit lies in holding the balance correctly between fire power and shock action . . . it must be accepted as a principle that the rifle, effective as it is, cannot replace the effect produced by the speed of the horse, the magnetism of the charge, and the terror of cold steel.[15]

The mood of the cavalryman on the eve of the First World War is perhaps best captured in an analysis of British military doctrine published in 1914:

Technically the great decisive cavalry charge on the main battlefield is a thing of the past, yet training in shock tactics is claimed by all cavalry authorities to be still essential to the strategic use of the arm, and even on the battlefield shock tactics may, under special conditions, conceivably still be possible, while brilliant opportunities will almost certainly be offered for the employment in perhaps a decisive manner of the power conferred by the combination of mobility with fire action. . . . For whatever tactics are adopted, the desire to take the offensive will always remain the breath of life for cavalry, and where shock action is impossible, the cavalryman must be prepared to expend, rifle in hand, the last man in an advance on foot, if the victory can thus only be achieved.[16]

So training in shock action continued; for even the reformers had to admit that cavalry would have to meet and defeat the enemy's cavalry, presumably in a gigantic mêlée, before it could fulfil its strategic task. "The opening of future wars," wrote von Bernhardi in 1912, "will, therefore, in all likelihood be characterised by great cavalry combats."[17]

So the cavalry continued to practice sword drill; and the infantry continued, for the same reason, to practice bayonet drill. The German writer Wilhelm Balck saw no reason to alter, in the 1911 edition of his huge study of *Tactics*, the doctrine preached in the first edition of 1896:

The soldier should be taught not to shrink from the bayonet attack, but to seek it. If the infantry is deprived of the arme blanche, if the impossibility of bayonet fighting is preached . . . an infantry will be developed which is unsuitable for attack and which moreover lacks a most essential quality, viz. the moral power to reach the enemy's position . . . [And he went on to quote from the Russian General Dragomirov, a well-known fanatic on the subject:] "The bayonet cannot be abolished for the reason, if for no other, that it is the sole and exclusive embodiment of that will-power which alone,

15. Quoted by Luvaas, *Military Legacy of the Civil War*, p. 107.
16. Maj. General E.A. Altham, *The Principles of War Historically Illustrated* (London, 1914), p. 92.
17. Bernhardi, *On War Today*, Vol. 2, p. 337.

both in war and in everyday life attains its object, whereas reason only facilitates the achievement of the object."[18]

The British General Staff manuals expressed the same idea slightly differently: "The moral effect of the bayonet is out of all proportion to its material effect, and not the least important of virtues claimed for it is that the desire to use it draws the attacking side on." To deprive the infantry of their bayonets would be like depriving the cavalry of their swords; it "would be to some extent to take away their desire to close."[19]

That brings us to the second point over which a rather more troubled consensus developed among European military thinkers as a consequence of the South African War: the unprecedented difficulty of carrying through frontal attacks, even with substantial artillery support, would now make necessary more extended formations in the attack. On this point also there had been a continuing controversy ever since 1870. The normal formation for the infantry attack, inherited from the Napoleonic era, consisted of three lines. First came the skirmishers in open formation, making maximum use of cover so as to reach positions from which they could bring a concentrated fire on the enemy in order, in cooperation with the artillery, to "win the fire fight." Behind them came the main assault line, normally in close formation under the immediate control of their officers, to assault with the bayonet. Finally came the supports, the immediate tactical reserve.

The German army, remembering the massacres of their infantry in the assault at the battles of Wörth and St. Privat in August 1870, had always inclined to the view that once the attacking infantry came under fire, close formations in the old style would be impossible. The main assault line would itself now have to scatter and edge its way forward to thicken up the skirmishers or extend their line, feeling for an exposed flank. Effectively it was now the skirmishers who bore the brunt of the attack, and success could be achieved only by the dominance of their fire. The bayonet, if used at all, would only gather up the harvest already reaped by the rifle and the gun.[20]

This was the doctrine against which Dragomirov and his disciples everywhere set their faces. It must be admitted that it did present real problems. Once the assaulting troops were scattered and left to themselves, out of range of the officers whose task it was to inspire them and the non-coms

18. Wilhelm Balck, *Tactics*, 4th ed. (Fort Leavenworth, 1911), Vol. 1, p. 383.
19. Altham, *Principles of War*, p. 80.
20. Balck, *Tactics*, Vol. 1, p. 373.

whose job it was to frighten them, what incentive would there be for them to go forward in face of enemy fire? Once they went to ground behind cover, would they ever get up again? There were several notorious instances in 1870 when substantial proportions of German assaulting formulations had unaccountably "got lost." Colonel Ardent du Picq, who had been killed in that war and whose posthumously published *Etudes sur le Combat* contain some of the shrewdest observations on troop morale that have ever been written, had described the terrifying isolation of the soldier on a modern battlefield (even before the days of smokeless powder) once he was deprived of the solid support of comrades on either side which had enabled men to face death ever since the days of the Roman legions. "The soldier is unknown even to his comrades; he loses them in the disorientating confusion of battle, where he fights as a lonely individual; solidarity is no longer guaranteed by mutual surveillance."[21] All now depended on the morale and reliability of the smallest units; "by force of circumstances all battles nowadays tend more than ever to become soldiers' battles."[22] How could these lonely frightened men, deprived of the intoxication of drums and trumpets, the support of their comrades, the inspiration of their leaders, find within themselves the courage to die?

The French army, its traditions of martial leadership and close formations for the attack antedating even the Napoleonic era, was particularly reluctant to accept the logic of the new firepower. For a decade after 1870 its leaders had attempted to impose the open tactical formations on their units, but they never really succeeded. By 1884 regulations were again prescribing "the principle of the decisive attack, head held high, unconcerned about casualties." The notorious regulations of 1894 laid it down that attacking units should advance elbow to elbow, not breaking formation to take advantage of cover, but assaulting *en masses* "to the sound of bugles and drums."[23] Stirring stuff, and the French were not alone in preferring it that way. So did the Russians, in spite of their chastening experiences before Plevna in 1877; and so did the British. They also, after a decade of uncertainty inspired by the events of 1870, returned to their old traditions. In the regulations of 1888, wrote Colonel Henderson:

---

21. Charles Ardent du Picq, *Etudes sur le Combat: Combat Antique et Moderne,* (Paris, repr. 1942), p. 110.
22. Ibid., p. 87.
23. Eugene Carrias, *La pensée militaire française* (Paris, 1960), p. 276.

> The bayonet has once more reasserted itself. To the second line, relying on cold steel only, as in the days of the Peninsula, is entrusted the duty of bringing the battle to a speedy conclusion. . . . The confusion of the Prussian battles was in a large degree due to their neglect of the immutable principles of tactics and . . . they are a bad model for us to follow. The sagacity of our own people is a surer guide and if, after 1870, we wanted a model, the tactics of the last great war waged by English-speaking soldiers would have served us better.

The Americans on both sides had always launched frontal attacks in close formations, having found that "to prevent the battle degenerating into a protracted struggle between two strongly entrenched armies, and to attain a speedy and decisive result, mere development of fire was insufficient." The lesson was clear: "close order whenever it is possible, extended order only when it is unavoidable."[24]

By 1900 Henderson was a sadder and a wiser man. Events in South Africa had once again shown the world that under fire close order was *not* possible; and the argument that it was good for morale was seen to be ludicrous. "When the preponderant mass suffers enormous losses; when they feel, as others will feel, that other and less costly means of achieving the same end might have been adopted, what will become of their morale? . . . The most brilliant offensive victories," went on Henderson, "are not those which were mere 'bludgeon work' and cost the most blood, but those which were won by surprise, by adroit manoeuvre, by mystifying and misleading the enemy, by turning the ground to the best account, and where the butchers' bill was small."[25] A generation later Henderson's countryman Liddell Hart was to elaborate this insight into an entire philosophy of war, but long before 1914 the British army was to discard this subversive suggestion that discretion might be the better part of valor.

Over the matter of close *versus* open formations for the attack, however, the South African experience was generally seen to be decisive. Even the French high command, while attributing the catastrophes which had overtaken the British entirely to Anglo–Saxon ineptitude, rewrote its regulations in 1904, abandoning the *coude à coude* formations of 1894 and prescribing advance by small groups covering each other by fire—the kind of infantry tactics that were to become general in the Second World War.[26] It is doubtful

24. Henderson, *Science of War*, pp. 135–150.
25. Ibid., pp. 373–375.
26. Carrias, *La pensée militaire française*, p. 290.

however whether these eminently sensible guidelines made any impression on an army which had been thrown, in the aftermath of the Dreyfus case, into a state of administrative confusion verging on anarchy.[27] Certainly the performance of the French infantry in 1914 shows no evidence of it. In any case, such tactics demanded of the ordinary soldier a degree of skill and self-reliance such as neither the French nor any other European army (with the possible exception of the Germans) had hitherto expected, or done anything to inculcate, either in their junior officers or in their other ranks.

And there remained unsolved the nagging, fundamental problem of *morale*—a problem all the greater since a large part of all armies would now be made up of reservists whose moral fiber, it was feared, would have been sapped by the enervating influences of civil life. Concern about the morale of the army was thus generalized, among European military thinkers, into concern about the morale of their nations as a whole; not so much whether they would stand up to the economic attrition which Bloch was almost unique in foreseeing, but whether they could inculcate into their young men that stoical contempt for death which alone would enable them to face, and overcome, the horrors of the assault.[28]

The Russo–Japanese War and the Superiority of the Offensive

It was while this concern was at its height that war broke out between Japan and Russia in the Far East. In February 1904 the Japanese navy launched a surprise attack on the Russian fleet at Port Arthur and, with local command of the sea thus secured, effected amphibious landings on the Korean and Manchurian coasts. It took the Japanese army a year to establish themselves in the disputed province of Manchuria, capturing Port Arthur by land assault and fighting its way north along the railway to capture the main Russian forward base at Mukden in a two-week battle involving altogether over half a million men. It was a war fought on both sides with the latest products of

---

27. Douglas Porch, *The March to the Marne* (Cambridge: Cambridge University Press, 1981), pp. 214–220.

28. "The steadily improving standards of living tend to increase the instinct of self-preservation and to diminish the spirit of self-sacrifice. . . . The fast manner of living at the present day undermines the nervous system, the fanaticism and religious and national enthusiasm of a bygone age are lacking, and finally the physical powers of the human species are also partly diminishing." Balck, *Tactics*, Vol. 1, p. 194. For equally gloomy British assessments see T.H.E. Travers, "Technology, Tactics, and Morale: Jean de Bloch, the Boer War, and British Military Theory 1900–1914," *Journal of Modern History*, Vol. 51, No. 2 (June 1979), pp. 264–286. This article is of seminal importance in showing the connection between tactical doctrine and national morale before 1914.

modern technology: not only magazine rifles and quick-firing field artillery but mobile heavy guns, machine guns, mines, barbed wire, searchlights, telephonic communications and, above all, *trenches*. The Russo–Japanese War proved beyond any doubt that the infantryman's most useful weapon, second only to his rifle, was a spade. Though the war inevitably had unique characteristics—both sides fought at the end of long supply lines, in sparsely inhabited country, which sharply limited the scale of force they could employ—it could not be dismissed, as so many conservative thinkers on the Continent dismissed the Boer War, as a colonial irrelevance. The Russian army was one of the greatest—certainly one of the largest—in Europe. The Japanese had had their armed forces equipped and trained by Europeans, mainly Germans, to the finest European standards. European—and American—military and naval observers with the fighting forces sent back expert reports on the operations, which were digested and mulled over by their general staffs. The British, the French, and the German armies all thought it worth their while to produce multi-volume histories of the Russo–Japanese War, and for the next ten years, until interest was eclipsed by events nearer home, its lessons were analyzed in the most precise detail by pundits writing in military periodicals. It was neither the Boer War nor the American Civil War nor even the Franco–Prussian War that European military specialists had in mind when their armies deployed in 1914: it was the fighting in Manchuria of 1904–5.

As usual, the experts tended to read into the experiences of the war very much what they wanted to find. Conservative cavalrymen observed the failure of the Russian cavalry, trained as it was to the use of the rifle, to achieve anything very much either on the battlefield or off it; absence of "the offensive spirit" making both its raids and its reconnaissance remarkably ineffectual. Reformers noted, on the contrary, how effectively the Japanese had deployed their cavalry in the role of mobile firepower, and the important part it had played at the battle of Mukden. Everyone agreed that artillery, with its accuracy, range, and rate of fire, was now of supreme importance; that it must almost always employ indirect fire; that shrapnel rather than high explosive was its most effective projectile; and that the consumption of ammunition would be enormous. Valuable lessons were learned about supply and communication problems and the need for inconspicuous uniforms; every European army quickly reclothed its armies in various shades of brown or grey, and it was political rather than military conservatism that fatally delayed this reform on the part of the French. But most important of all was the general consensus that infantry assaults with the bayonet, in spite of the South African experience, were still not only possible but necessary. The Japanese had carried them out time and again, and usually with ultimate success.

The Japanese bayonet assaults came, it was true, only at the end of a long and careful advance. They approached whenever possible by night, digging in before dawn, lying up by day, and repeating the process until they could get no further. Then, breaking completely with the European tradition of advancing in extended lines, they dashed forward in small groups of one or two dozen men, each with its own objective, moving rapidly from cover to cover until they were sufficiently close to assault. A French observer described one such scene:

> The whole Japanese line is now lit up with the glitter of steel flashing from the scabbard. . . . Once again the officers quit shelter with ringing shouts of "Banzai!" wildly echoed by all the rank and file. Slowly, but not to be denied, they make headway, in spite of the barbed wire, mines and pitfalls, and the merciless hail of bullets. Whole units are destroyed—others take their places; the advancing wave pauses for a moment, but sweeps ever onward. Already they are within a few yards of the trenches. Then, on the Russian side, the long grey line of Siberian Fusiliers forms up in turn, and delivers one last volley before scurrying down the far side of the hill at the double.[29]

The Japanese losses in these assaults were heavy, but they succeeded; and, so argued the European theorists, such tactics would succeed again. "The Manchurian experience," as one British military writer put it, "showed over and over again that the bayonet was in no sense an obsolete weapon. . . . The assault is even of more importance than the attainment of fire mastery which antecedes it. It is the supreme moment of the fight. . . . Upon it the final issue depends. . . . From these glorious examples it may be deduced that no duty, however difficult, should be regarded as impossible by well-trained infantry of good morale and discipline."[30]

It was this "morale and discipline" of the Japanese armed forces that all observers stressed, and they were equally unanimous in stressing that these qualities characterized not only the armed forces but the entire Japanese nation. General Kuropatkin, the commander of the Russian forces, noted ruefully in his memoirs:

29. General François de Négrier, *Lessons from the Russo-Japanese War* (London, 1905), p. 69.
30. Altham, *Principles of War*, pp. 295–6, 302.

In the late war . . . our moral strength was less than that of the Japanese; and it was this inferiority, rather than mistakes in generalship, that caused our defeats. . . . The lack of martial spirit, of moral exaltation, and of heroic impulse, affected particularly our stubbornness in battle. In many cases we did not have sufficient resolution to conquer such antagonists as the Japanese.[31]

The same quality gave a representative of Japan's British ally, General Sir Ian Hamilton, almost equal concern:

It is not so much the idea that we have put our money on the wrong horse that now troubles me. . . . But it should cause European statesmen some anxiety when their people seem to forget that there are millions outside the charmed circle of Western Civilisation who are ready to pluck the sceptre from nerveless hands so soon as the old spirit is allowed to degenerate. . . . Providentially Japan is our ally. . . . England has time, therefore—time to put her military affairs in order; time to implant and cherish the military ideal in the hearts of her children; time to prepare for a disturbed and an anxious twentieth century. . . . From the nursery and its toys to the Sunday school and its cadet company, every influence of affection, loyalty, tradition and education should be brought to bear on the next generation of British boys and girls, so as deeply to impress upon their young minds a feeling of reverence and admiration for the patriotic spirit of their ancestors.[32]

Such expressions of admiration for the creed of Bushido are to be found widely scattered in the military and militarist literature of the day. Particularly important for our purposes, however, was the general recognition that the Japanese performance had proved, up to the hilt, the moral and military superiority of *the offensive*. The passive immobility of the Russians, in spite of all the advantages they should have enjoyed from the defense, had in the long run ensured their defeat. It was a conclusion which the military everywhere, after the miasmic doubts engendered by the Boer War, embraced with heartfelt relief. "The defensive is never an acceptable role to the Briton, and he makes little or no study of it," wrote Major General Sir W.G. Knox flatly in 1914.[33] "It was not by dwelling on the idea of passive defense," wrote the Secretary of State for War R.B. Haldane in 1911, "that our fore-

31. General G.N. Kuropatkin, *The Russian Army and the Japanese War* (London, 1909), Vol. 2, p. 80.
32. Major General Sir Ian Hamilton, *A Staff Officer's Scrapbook* (London, 1905), Vol. 1, pp. 10–13.
33. Quoted by Travers, "Technology, Tactics, and Morale."

fathers made our country what it is today."[34] In Germany General von Schlieffen, on retiring as Chief of the General Staff in 1905, held up to his successors the model of the German armies in 1870: "Attacks, and more attacks, ruthless attacks brought it unparalleled losses but also victory and, it is probably true to say, the decision of the campaign."[35] And his successor, the younger von Moltke, acknowledged the heritage: "We have learned the object that you seek to achieve: not to obtain limited successes but to strike great, destructive blows. . . . Your object is the annihilation of the enemy, and all efforts must be directed towards this end."[36]

Nowhere was the lesson more gratefully received, however, than in France. Marshal Joffre, whose offensive operations from 1914 through 1916 are now generally considered to have been a succession of unmitigated disasters, described the French reaction to the Russo–Japanese War in his Memoirs with quite unrepentant frankness. After the Boer War, he wrote,

> a whole series of false doctrines . . . began to undermine even such feeble offensive sentiment as had made its appearance in our war doctrines . . . an incomplete study of the events of a single war had led the intellectual elite of our Army to believe that the improvement in firearms and the power of fire action had so increased the strength of the defensive that an offensive opposed to it had lost all virtue.

After the Russo–Japanese War, however,

> our young intellectual elite finally shook off the malady of this phraseology which had upset the military world and returned to a more healthy conception of the general conditions prevailing in war.[37]

Joffre admitted that the new passion for the offensive did take on a "somewhat unreasoning character," citing Colonel de Grandmaison's famous lectures of 1911 as an example. "Unreasoning" is the right word. One must always, declared de Grandmaison to his audience,

> succeed in combat in doing things which would be *impossible* in cold blood. For instance . . . advancing under fire. . . . We must prepare ourselves for

34. R.B. Haldane, Introduction to Sir Ian Hamilton, *Compulsory Service*, 2nd ed. (London, 1911), p. 38.
35. Quoted by Fritz Fischer, *War of Illusions* (New York and London: W. W. Norton, 1975), p. 395.
36. Eugene Carrias, *La pensée militaire allemande* (Paris, 1948), p. 319.
37. Joseph Joffre, *The Personal Memoirs of Marshal Joffre* (London: Harper & Brothers, 1932), Vol. 1, pp. 27ff.

it, and prepare others by cultivating, passionately, everything which bears the mark of the offensive spirit. To take this to excess would probably still not be far enough.[38]

There was nothing in this to indicate the careful use of ground and of mutual fire support which had characterized the actual Japanese tactics—tactics in fact remarkably close to those prescribed in the despised French infantry regulation of 1904. But de Grandmaison was not so much setting out a military doctrine as echoing a national mood—a generalized sense of chauvinistic assertiveness which dominated the French "establishment," civil and military alike, in 1911–12.[39] It was a mood which did much to restore the morale of an army battered and confused after the excesses of the Dreyfus affair, but it could not of itself create the battlefield skills which had also characterized the Japanese army, and without which "the spirit of the offensive" was not so much an assertion of national morale as a generalized death wish. It was in this mood that French officers led the attacks in August–September 1914 which within six weeks produced 385,000 casualties, of which 100,000 were dead.[40]

Bloch died in 1902, but he could have taken much comfort from the experiences of the Russo–Japanese War. Its battles were prolonged, costly, and indecisive. Victory came through attrition; and defeat, for Russia, brought revolution. But Bloch's critics could equally well argue that his major thesis had been disproved. War had been shown to be neither impossible, nor suicidal. It was still a highly effective instrument of policy for a nation which had the courage to face its dangers and the endurance to bear its costs—especially its inevitable and predictable costs in human lives. Those nations which were not prepared to put their destinies to this test, they urged, could expect no mercy in the grim battle for survival which had always characterized human history and which seemed likely, in the coming century, to be waged with ever greater ferocity. It was in this mood, and with these hopes, that the nations of Europe went to war in 1914.

---

38. Quoted in Henri Contamine, *La Revanche 1871–1914* (Paris, 1957), p. 167.
39. Eugene Weber, *The Nationalist Revival in France, 1905–1914* (Berkeley and Los Angeles: University of California Press, 1959), pp. 93–105.
40. Contamine, *La Revanche*, p. 276.

The Cult of the Offensive and the Origins of the First World War

Stephen Van Evera

During the decades before the First World War a phenomenon which may be called a "cult of the offensive" swept through Europe. Militaries glorified the offensive and adopted offensive military doctrines, while civilian elites and publics assumed that the offense had the advantage in warfare, and that offensive solutions to security problems were the most effective.

This article will argue that the cult of the offensive was a principal cause of the First World War, creating or magnifying many of the dangers which historians blame for causing the July crisis and rendering it uncontrollable. The following section will first outline the growth of the cult of the offensive in Europe in the years before the war, and then sketch the consequences which international relations theory suggests should follow from it. The second section will outline consequences which the cult produced in 1914, and the final section will suggest conclusions and implications for current American policy.

The Cult of the Offensive and International Relations Theory

THE GROWTH OF THE CULT

The gulf between myth and the realities of warfare has never been greater than in the years before World War I. Despite the large and growing advantage which defenders gained against attackers as a result of the invention of rifled and repeating small arms, the machine gun, barbed wire, and the development of railroads, Europeans increasingly believed that attackers would hold the advantage on the battlefield, and that wars would be short and "decisive"—a "brief storm," in the words of the German Chancellor,

I would like to thank Jack Snyder, Richard Ned Lebow, Barry Posen, Marc Trachtenberg, and Stephen Walt for their thoughtful comments on earlier drafts of this paper.

Stephen Van Evera is a Research Fellow at the Center for Science and International Affairs, Harvard University.

International Security, Summer 1984 (Vol. 9, No. 1) 0162-2889/84/010058-50 $02.50/1

Bethmann Hollweg.[1] They largely overlooked the lessons of the American Civil War, the Russo–Turkish War of 1877–78, the Boer War, and the Russo–Japanese War, which had demonstrated the power of the new defensive technologies. Instead, Europeans embraced a set of political and military myths which obscured both the defender's advantages and the obstacles an aggressor would confront. This mindset helped to mold the offensive military doctrines which every European power adopted during the period 1892–1913.[2]

In Germany, the military glorified the offense in strident terms, and inculcated German society with similar views. General Alfred von Schlieffen, author of the 1914 German war plan, declared that "Attack is the best defense," while the popular publicist Friedrich von Bernhardi proclaimed that "the offensive mode of action is by far superior to the defensive mode," and that "the superiority of offensive warfare under modern conditions is greater than formerly."[3] German Chief of Staff General Helmuth von Moltke also endorsed "the principle that the offensive is the best defense," while General August von Keim, founder of the Army League, argued that "Germany ought to be armed for attack," since "the offensive is the only way of insuring victory."[4] These assumptions guided the Schlieffen Plan, which envisaged rapid and decisive attacks on Belgium, France, and Russia.

---

1. Quoted in L.L. Farrar, Jr., "The Short War Illusion: The Syndrome of German Strategy, August–December 1914," *Militaergeschictliche Mitteilungen*, No. 2 (1972), p. 40.
2. On the origins of the cult of the offensive, see Jack Lewis Snyder, "Defending the Offensive: Biases in French, German, and Russian War Planning, 1870–1914" (Ph.D. dissertation, Columbia University, 1981), forthcoming as a book from Cornell University Press in 1984; Snyder's essay in this issue; and my "Causes of War" (Ph.D. dissertation, University of California, Berkeley, 1984), chapter 7. On the failure of Europeans to learn defensive lessons from the wars of 1860–1914, see Jay Luvaas, *The Military Legacy of the Civil War: The European Inheritance* (Chicago: University of Chicago Press, 1959); and T.H.E. Travers, "Technology, Tactics, and Morale: Jean de Bloch, the Boer War, and British Military Theory, 1900–1914," *Journal of Modern History*, Vol. 51 (June 1979), pp. 264–286. Also relevant is Bernard Brodie, *Strategy in the Missile Age* (Princeton: Princeton University Press, 1965), pp. 42–52.

A related work which explores the sources of offensive and defensive doctrines before World War II is Barry R. Posen, *The Sources of Military Doctrine: France, Britain, and Germany Between the World Wars* (Ithaca: Cornell University Press, 1984), pp. 47–51, 67–74, and passim.

3. Gerhard Ritter, *The Schlieffen Plan: Critique of a Myth*, trans. Andrew and Eva Wilson, with a Foreword by B.H. Liddell Hart (London: Oswald Wolff, 1958; reprint ed., Westport, Conn.: Greenwood Press, 1979), p. 100; and Friedrich von Bernhardi, *How Germany Makes War* (New York: George H. Doran Co., 1914), pp. 153, 155.
4. Imanuel Geiss, ed., *July 1914: The Outbreak of the First World War: Selected Documents* (New York: W.W. Norton, 1967), p. 357; and Wallace Notestein and Elmer E. Stoll, eds., *Conquest and Kultur: Aims of the Germans in Their Own Words* (Washington, D.C.: U.S. Government Printing Office, 1917), p. 43. Similar ideas developed in the German navy; see Holger H. Herwig, *Politics*

In France, the army became "Obsessed with the virtues of the offensive," in the words of B.H. Liddell Hart, an obsession which also spread to French civilians.[5] The French army, declared Chief of Staff Joffre, "no longer knows any other law than the offensive. . . . Any other conception ought to be rejected as contrary to the very nature of war,"[6] while the President of the French Republic, Clément Fallières, announced that "The offensive alone is suited to the temperament of French soldiers. . . . We are determined to march straight against the enemy without hesitation."[7] Emile Driant, a member of the French chamber of deputies, summarized the common view: "The first great battle will decide the whole war, and wars will be short. The idea of the offense must penetrate the spirit of our nation."[8] French military doctrine reflected these offensive biases.[9] In Marshall Foch's words, the French army adopted "a single formula for success, a single combat doctrine, namely, the decisive power of offensive action undertaken with the resolute determination to march on the enemy, reach and destroy him."[10]

Other European states displayed milder symptoms of the same virus. The British military resolutely rejected defensive strategies despite their experience in the Boer War which demonstrated the power of entrenched defenders against exposed attackers. General W.G. Knox wrote, "The defensive is never an acceptable role to the Briton, and he makes little or no study of it," and General R.C.B. Haking argued that the offensive "will win as sure as there is a sun in the heavens."[11] The Russian Minister of War, General V.A. Sukhomlinov, observed that Russia's enemies were directing their armies "towards guaranteeing the possibility of dealing rapid and decisive blows.

---

of Frustration: The United States in German Naval Planning, 1889–1941 (Boston: Little, Brown & Co., 1976), pp. 42–66.

5. B.H. Liddell Hart, *Through the Fog of War* (New York: Random House, 1938), p. 57.

6. In 1912, quoted in John Ellis, *The Social History of the Machine Gun* (New York: Pantheon, 1975), pp. 53–54.

7. Barbara Tuchman, *The Guns of August* (New York: Dell, 1962), p. 51.

8. In 1912, quoted in John M. Cairns, "International Politics and the Military Mind: The Case of the French Republic, 1911–1914," *The Journal of Modern History*, Vol. 25, No. 3 (September 1953), p. 282.

9. On the offensive in French prewar thought, see B.H. Liddell Hart, "French Military Ideas before the First World War," in Martin Gilbert, ed., *A Century of Conflict, 1850–1950* (London: Hamilton Hamish, 1966), pp. 135–148.

10. Richard D. Challener, *The French Theory of the Nation in Arms, 1866–1939* (New York: Columbia University Press, 1955), p. 81. Likewise, Joffre later explained that Plan XVII, his battle plan for 1914, was less a plan for battle than merely a plan of "concentration. . . . I adopted no preconceived idea, other than a full determination to take the offensive with all my forces assembled." Theodore Ropp, *War in the Modern World*, rev. ed. (New York: Collier, 1962), p. 229.

11. In 1913 and 1914, quoted in Travers, "Technology, Tactics, and Morale," p. 275.

. . . We also must follow this example."[12] Even in Belgium the offensive found proponents: under the influence of French ideas, some Belgian officers favored an offensive strategy, proposing the remarkable argument that "To ensure against our being ignored it was essential that we should attack," and declaring that "We must hit them where it hurts."[13]

Mythical or mystical arguments obscured the technical dominion of the defense, giving this faith in the offense aspects of a cult, or a mystique, as Marshall Joffre remarked in his memoirs.[14] For instance, Foch mistakenly argued that the machine gun actually strengthened the offense: "Any improvement of firearms is ultimately bound to add strength to the offensive. . . . Nothing is easier than to give a mathematical demonstration of that truth." If two thousand men attacked one thousand, each man in both groups firing his rifle once a minute, he explained, the "balance in favor of the attack" was one thousand bullets per minute. But if both sides could fire ten times per minute, the "balance in favor of the attacker" would increase to ten thousand, giving the attack the overall advantage.[15] With equally forced logic, Bernhardi wrote that the larger the army the longer defensive measures would take to execute, owing to "the difficulty of moving masses"; hence, he argued, as armies grew, so would the relative power of the offense.[16]

British and French officers suggested that superior morale on the attacking side could overcome superior defensive firepower, and that this superiority in morale could be achieved simply by assuming the role of attacker, since offense was a morale-building activity. One French officer contended that "the offensive doubles the energy of the troops" and "concentrates the thoughts of the commander on a single objective,"[17] while British officers declared that "Modern [war] conditions have enormously increased the value of moral quality," and "the moral attributes [are] the primary causes of all great success."[18] In short, mind would prevail over matter; morale would triumph over machine guns.

---

12. In 1909, quoted in D.C.B. Lieven, *Russia and the Origins of the First World War* (New York: St. Martin's Press, 1983), p. 113.
13. See Tuchman, *Guns of August*, pp. 127–131.
14. Marshall Joffre, *Mémoires du Maréchel Joffre* (Paris: Librarie Plon, 1932), p. 33. Joffre speaks of "le culte de l'offensive" and "d'une 'mystique de l'offensive'" of "le caractère un peu irraisonné."
15. Ropp, *War in the Modern World*, p. 218.
16. Ibid., p. 203. See also Bernhardi, *How Germany Makes War*, p. 154.
17. Captain Georges Gilbert, quoted in Snyder, "Defending the Offensive," pp. 80–81.
18. The *Field Service Regulations* of 1909 and Colonel Kiggell, quoted in Travers, "Technology, Tactics, and Morale," pp. 273, 276–277.

Even when European officers recognized the new tactical power of the defense, they often

Europeans also tended to discount the power of political factors which would favor defenders. Many Germans believed that "bandwagoning" with a powerful state rather than "balancing" against it was the guiding principle in international alliance-formation.[19] Aggressors would gather momentum as they gained power, because opponents would be intimidated into acquiescence and neutrals would rally to the stronger side. Such thinking led German Chancellor Bethmann Hollweg to hope that "Germany's growing strength . . . might force England to realize that [the balance of power] principle had become untenable and impracticable and to opt for a peaceful settlement with Germany,"[20] and German Secretary of State Gottlieb von Jagow to forecast British neutrality in a future European war: "We have not built our fleet in vain," and "people in England will seriously ask themselves whether it will be just that simple and without danger to play the role of France's guardian angel against us."[21] German leaders also thought they might frighten Belgium into surrender: during the July crisis Moltke was "counting on the possibility of being able to come to an understanding [with Belgium] when the Belgian Government realizes the seriousness of the situation."[22] This ill-founded belief in bandwagoning reinforced the general belief that conquest was relatively easy.

The belief in easy conquest eventually pervaded public images of international politics, manifesting itself most prominently in the widespread application of Darwinist notions to international affairs. In this image, states competed in a decisive struggle for survival which weeded out the weak and ended in the triumph of stronger states and races—an image which assumed a powerful offense. "In the struggle between nationalities," wrote former

---

resisted the conclusion that the defender would also hold the strategic advantage. Thus Bernhardi wrote that while "the defense as a form of fighting is stronger than the attack," it remained true that "in the conduct of war as a whole the offensive mode of action is by far superior to the defensive mode, especially under modern conditions." Bernhardi, *How Germany Makes War,* p. 155. See also Snyder, "Defending the Offensive," pp. 152–154, 253–254; and Travers, "Technology, Tactics, and Morale," passim.

19. On these concepts, see Kenneth N. Waltz, *Theory of International Politics* (Reading, Mass.: Addison–Wesley, 1979), pp. 125–127; and Stephen M. Walt, "The Origins of Alliances" (Ph.D. dissertation, University of California, Berkeley, 1983).

20. December 2, 1914, quoted in Fritz Fischer, *War of Illusions: German Policies from 1911 to 1914,* trans. Marian Jackson, with a Foreword by Alan Bullock (New York: W.W. Norton, 1975), p. 69.

21. February 1914, quoted in Geiss, *July 1914,* p. 25. For more examples, see Fischer, *War of Illusions,* pp. 133, 227; and Wayne C. Thompson, *In the Eye of the Storm: Kurt Riezler and the Crises of Modern Germany* (Iowa City: University of Iowa Press, 1980), p. 120.

22. August 3, quoted in Bernadotte E. Schmitt, *The Coming of the War: 1914,* 2 vols. (New York: Charles Scribner's Sons, 1930), Vol. 2, p. 390n.

German Chancellor Bernhard von Bülow, "one nation is the hammer and the other the anvil; one is the victor and the other the vanquished. . . . it is a law of life and development in history that where two national civilisations meet they fight for ascendancy."[23] A writer in the London *Saturday Review* portrayed the Anglo–German competition as "the first great racial struggle of the future: here are two growing nations pressing against each other . . . all over the world. One or the other has to go; one or the other will go."[24] This Darwinist foreign policy thought reflected and rested upon the implicit assumption that the offense was strong, since "grow or die" dynamics would be impeded in a defense-dominant world where growth could be stopped and death prevented by self-defense.

CONSEQUENCES OF OFFENSE-DOMINANCE

Recent theoretical writing in international relations emphasizes the dangers that arise when the offense is strong relative to the defense.[25] If the theory outlined in these writings is valid, it follows that the cult of the offensive was a reason for the outbreak of the war.

Five major dangers relevant to the 1914 case may develop when the offense is strong, according to this recent writing. First, states adopt more aggressive

---

23. Prince Bernhard von Bülow, *Imperial Germany*, trans. Marie A. Lewenz (New York: Dodd, Mead & Co., 1915), p. 291. On international social Darwinism, see also H.W. Koch, "Social Imperialism as a Factor in the 'New Imperialism,'" in H.W. Koch, ed., *The Origins of the First World War* (London: Macmillan, 1972), pp. 329–354.
24. Joachim Remak, *The Origins of World War I, 1871–1914* (Hinsdale, Ill.: Dryden Press, 1967), p. 85. Likewise the British Colonial Secretary, Joseph Chamberlain, declared that "the tendency of the time is to throw all power into the hands of the greater empires," while the "minor kingdoms" seemed "destined to fall into a secondary and subordinate place. . . ." In 1897, quoted in Fischer, *War of Illusions*, p. 35.
25. See Robert Jervis's pathbreaking article, "Cooperation under the Security Dilemma," *World Politics*, Vol. 30, No. 2 (January 1978), pp. 167–214; and Chapter 3 of my "Causes of War." Also relevant are George H. Quester, *Offense and Defense in the International System* (New York: John Wiley & Sons, 1977); John Herz, "Idealist Internationalism and the Security Dilemma," *World Politics*, Vol. 2, No. 2 (January 1950), pp. 157, 163; and Herbert Butterfield, *History and Human Relations* (London: Collins, 1950), pp. 19–20. Applications and elaborations include: Shai Feldman, *Israeli Nuclear Deterrence* (New York: Columbia University Press, 1982); idem, "Superpower Security Guarantees in the 1980's," in *Third World Conflict and International Security, Part II*, Adelphi Paper No. 167 (London: International Institute for Strategic Studies, 1981), pp. 34–44; Barry R. Posen, "Inadvertent Nuclear War? Escalation and NATO's Northern Flank," *International Security*, Vol. 7, No. 2 (Fall 1982), pp. 28–54; Jack Lewis Snyder, "Perceptions of the Security Dilemma in 1914," in Robert Jervis and Richard Ned Lebow, eds., *Perceptions and Deterrence*, forthcoming in 1985; and Kenneth N. Waltz, *The Spread of Nuclear Weapons: More May Be Better*, Adelphi Paper No. 171 (London: International Institute for Strategic Studies, 1981). Of related interest is John J. Mearsheimer, *Conventional Deterrence* (Ithaca: Cornell University Press, 1983).

foreign policies, both to exploit new opportunities and to avert new dangers which appear when the offense is strong. Expansion is more tempting, because the cost of aggression declines when the offense has the advantage. States are also driven to expand by the need to control assets and create the conditions they require to secure themselves against aggressors, because security becomes a scarcer asset. Alliances widen and tighten as states grow more dependent on one another for security, a circumstance which fosters the spreading of local conflicts. Moreover, each state is more likely to be menaced by aggressive neighbors who are governed by the same logic, creating an even more competitive atmosphere and giving states further reason to seek security in alliances and expansion.

Second, the size of the advantage accruing to the side mobilizing or striking first increases, raising the risk of preemptive war.[26] When the offense is strong, smaller shifts in ratios of forces between states create greater shifts in their relative capacity to conquer territory. As a result states have greater incentive to mobilize first or strike first, if they can change the force ratio in their favor by doing so. This incentive leads states to mobilize or attack to

26. In a "preemptive" war, either side gains by moving first; hence, one side moves to exploit the advantage of moving first, or to prevent the other side from doing so. By contrast, in a "preventive" war, one side foresees an adverse shift in the balance of power, and attacks to avoid a more difficult fight later.

"Moving first" in a preemptive war can consist of striking first *or mobilizing* first, if mobilization sets in train events which cause war, as in 1914. Thus a war is preemptive if statesmen attack because they believe that it pays to strike first; or if they mobilize because they believe that it pays to mobilize first, even if they do not also believe that it pays to strike first, if mobilizations open "windows" which spur attacks for "preventive" reasons, or if they produce other effects which cause war. Under such circumstances war is caused by preemptive actions which are not acts of war, but which are their equivalent since they produce conditions which cause war.

A preemptive war could also involve an attack by one side and mobilization by the other—for instance, one side might mobilize to forestall an attack, or might attack to forestall a mobilization, as the Germans apparently attacked Liège to forestall Belgian preparations to defend it (see below). Thus four classes of preemption are possible: an attack to forestall an attack, an attack to forestall a mobilization, a mobilization to forestall an attack, or a mobilization to forestall a mobilization (such as the Russian mobilizations in 1914).

The size of the incentive to preempt is a function of three factors: the degree of secrecy with which each side could mobilize its forces or mount an attack; the change in the ratio of forces which a secret mobilization or attack would produce; and the size and value of the additional territory which this changed ratio would allow the attacker to conquer or defend. If secret action is impossible, or if it would not change force ratios in favor of the side moving first, or if changes in force ratios would not change relative ability to conquer territory, then there is no first-strike or first-mobilization advantage. Otherwise, states have some inducement to move first.

On preemption, see Thomas C. Schelling, *Arms and Influence* (New Haven: Yale University Press, 1966), pp. 221–259; and idem, *Strategy of Conflict* (New York: Oxford University Press, 1963), pp. 207–254.

seize the initiative or deny it to adversaries, and to conceal plans, demands, and grievances to avoid setting off such a strike by their enemies, with deleterious effects on diplomacy.

Third, "windows" of opportunity and vulnerability open wider, forcing faster diplomacy and raising the risk of preventive war. Since smaller shifts in force ratios have larger effects on relative capacity to conquer territory, smaller prospective shifts in force ratios cause greater hope and alarm, open bigger windows of opportunity and vulnerability, and enhance the attractiveness of exploiting a window by launching a preventive attack.

Fourth, states adopt more competitive styles of diplomacy—brinkmanship and presenting opponents with *faits accomplis,* for instance—since the gains promised by such tactics can more easily justify the risks they entail. At the same time, however, the risks of adopting such strategies also increase, because they tend to threaten the vital interests of other states more directly. Because the security of states is more precarious and more tightly interdependent, threatening actions force stronger and faster reactions, and the political ripple effects of *faits accomplis* are larger and harder to control.

Fifth, states enforce tighter political and military secrecy, since national security is threatened more directly if enemies win the contest for information. As with all security assets, the marginal utility of information is magnified when the offense is strong; hence states compete harder to gain the advantage and avoid the disadvantage of disclosure, leading states to conceal their political and military planning and decision-making more carefully.

The following section suggests that many of the proximate causes of the war of 1914 represent various guises of these consequences of offense-dominance: either they were generated or exacerbated by the assumption that the offense was strong, or their effects were rendered more dangerous by this assumption. These causes include: German and Austrian expansionism; the belief that the side which mobilized or struck first would have the advantage; the German and Austrian belief that they faced "windows of vulnerability"; the nature and inflexibility of the Russian and German war plans and the tight nature of the European alliance system, both of which spread the war from the Balkans to the rest of Europe; the imperative that "mobilization meant war" for Germany; the failure of Britain to take effective measures to deter Germany; the uncommon number of blunders and mistakes committed by statesmen during the July crisis; and the ability of the Central powers to evade blame for the war. Without the cult of the offensive these problems probably would have been less acute, and their effects would have posed smaller risks. Thus the cult of the offensive was a mainspring driving many of the mechanisms which brought about the First World War.

The Cult of the Offensive and the Causes of the War

GERMAN EXPANSION AND ENTENTE RESISTANCE

Before 1914 Germany sought a wider sphere of influence or empire, and the war grew largely from the political collision between expansionist Germany and a resistant Europe. Germans differed on whether their empire should be formal or informal, whether they should seek it in Europe or overseas, and whether they should try to acquire it peacefully or by violence, but a broad consensus favored expansion of some kind. The logic behind this expansionism, in turn, rested on two widespread beliefs which reflected the cult of the offensive: first, that German security required a wider empire; and second, that such an empire was readily attainable, either by coercion or conquest. Thus German expansionism reflected the assumption that conquest would be easy both for Germany and for its enemies.

Prewar statements by German leaders and intellectuals reflected a pervasive belief that German independence was threatened unless Germany won changes in the status quo. Kaiser Wilhelm foresaw a "battle of Germans against the Russo–Gauls for their very existence," which would decide "the existence or non-existence of the Germanic race in Europe,"[27] declaring: "The question for Germany is to be or not to be."[28] His Chancellor, Bethmann Hollweg, wondered aloud if there were any purpose in planting new trees at his estate at Hohenfinow, near Berlin, since "in a few years the Russians would be here anyway."[29] The historian Heinrich von Treitschke forecast that "in the long run the small states of central Europe can not maintain themselves,"[30] while other Germans warned, "If Germany does not rule the world . . . it will disappear from the map; it is a question of either or," and "Germany will be a world power or nothing."[31] Similarly, German military officers predicted that "without colonial possessions [Germany] will suffocate in her small territory or else will be crushed by the great world powers" and

---

27. In 1912, quoted in Thompson, *Eye of the Storm*, p. 42.
28. In 1912, quoted in Fischer, *War of Illusions*, p. 161.
29. V.R. Berghahn, *Germany and the Approach of War in 1914* (London: Macmillan, 1973), p. 186.
30. In 1897, quoted in Notestein and Stoll, *Conquest and Kultur*, p. 21.
31. Houston Chamberlain and Ernest Hasse, quoted in Fischer, *War of Illusions*, pp. 30, 36.

foresaw a "supreme struggle, in which the existence of Germany will be at stake. . . ."[32]

Germans also widely believed that expansion could solve their insecurity: "Room; they must make room. The western and southern Slavs—or we! . . . Only by growth can a people save itself."[33] German expansionists complained that German borders were constricted and indefensible, picturing a Germany "badly protected by its unfavorable geographic frontiers. . . ."[34] Expansion was the suggested remedy: "Our frontiers are too narrow. We must become land-hungry, must acquire new regions for settlement. . . ."[35] Expanded borders would provide more defensible frontiers and new areas for settlement and economic growth, which in turn would strengthen the German race against its competitors: "the continental expansion of German territory [and] the multiplication on the continent of the German peasantry . . . would form a sure barrier against the advance of our enemies. . . ."[36] Such utterances came chiefly from the hawkish end of the German political spectrum, but they reflected widely held assumptions.

Many Germans also failed to see the military and political obstacles to expansion. The Kaiser told departing troops in early August, "You will be home before the leaves have fallen from the trees,"[37] and one of his generals predicted that the German army would sweep through Europe like a bus full of tourists: "In two weeks we shall defeat France, then we shall turn round, defeat Russia and then we shall march to the Balkans and establish order there."[38] During the July crisis a British observer noted the mood of "supreme confidence" in Berlin military circles, and a German observer reported that the German General Staff "looks ahead to war with France with great confidence, expects to defeat France within four weeks. . . ."[39] While some

---

32. *Nauticus*, in 1900, quoted in Berghahn, *Germany and the Approach of War in 1914*, p. 29; and Colmar von der Goltz, quoted in Notestein and Stoll, *Conquest and Kultur*, p. 119.

33. Otto Richard Tannenberg, in 1911, quoted in Notestein and Stoll, *Conquest and Kultur*, p. 53.

34. Crown Prince Wilhelm, in 1913, quoted in ibid., p. 44. Likewise Walter Rathenau complained of German "frontiers which are too long and devoid of natural protection, surrounded and hemmed in by rivals, with a short coastline. . . ." In July 1914, quoted in Fischer, *War of Illusions*, p. 450.

35. Hermann Vietinghoff-Scheel, in 1912, quoted in William Archer, ed., *501 Gems of German Thought* (London: T. Fisher Unwin, 1916), p. 46.

36. Albrecht Wirth, in 1901, quoted in Notestein and Stoll, *Conquest and Kultur*, p. 52.

37. Quoted in Tuchman, *Guns of August*, p. 142.

38. Von Loebell, quoted in Fischer, *War of Illusions*, p. 543.

39. The English Military Attaché, quoted in Luigi Albertini, *The Origins of the War of 1914*, 3 vols., trans. and ed. Isabella M. Massey (London: Oxford University Press, 1952–57; reprint ed.,

German military planners recognized the tactical advantage which defenders would hold on the battlefield, most German officers and civilians believed they could win a spectacular, decisive victory if they struck at the right moment.

Bandwagon logic fed hopes that British and Belgian opposition to German expansion could be overcome. General Moltke believed that "Britain is peace loving" because in an Anglo–German war "Britain will lose its domination at sea which will pass forever to America"[40]; hence Britain would be intimidated into neutrality. Furthermore, he warned the Belgians, "Small countries, such as Belgium, would be well advised to rally to the side of the strong if they wished to retain their independence," expecting Belgium to follow this advice if Germany applied enough pressure.[41]

Victory, moreover, would be decisive and final. In Bülow's words, a defeat could render Russia "incapable of attacking us for at least a generation" and "unable to stand up for twenty-five years," leaving it "lastingly weakened,"[42] while Bernhardi proposed that France "must be annihilated once and for all as a great power."[43]

Thus, as Robert Jervis notes: "Because of the perceived advantage of the offense, war was seen as the best route both to gaining expansion and to avoiding drastic loss of influence. There seemed to be no way for Germany merely to retain and safeguard her existing position."[44] The presumed power of the offense made empire appear both feasible and necessary. Had Germans recognized the real power of the defense, the notion of gaining wider empire would have lost both its urgency and its plausibility.

Security was not Germany's only concern, nor was it always a genuine one. In Germany, as elsewhere, security sometimes served as a pretext for expansion undertaken for other reasons. Thus proponents of the "social imperialism" theory of German expansion note that German elites endorsed imperialism, often using security arguments, partly to strengthen their do-

---

Westport, Conn.: Greenwood Press, 1980), Vol. 3, p. 171; and Lerchenfeld, the Bavarian ambassador in Berlin, quoted in Fischer, *War of Illusions*, p. 503.
40. In 1913, quoted in Fischer, *War of Illusions*, p. 227.
41. In 1913, quoted in Albertini, *Origins of the War*, Vol. 3, p. 441. See also Bernhardi's dismissal of the balance of power, in Friedrich von Bernhardi, *Germany and the Next War*, trans. Allen H. Powles (New York: Longmans, Green & Co., 1914), p. 21.
42. In 1887, quoted in Fischer, *War of Illusions*, p. 45.
43. In 1911, quoted in Tuchman, *Guns of August*, p. 26.
44. Jervis, "Cooperation under the Security Dilemma," p. 191.

mestic political and social position.[45] Likewise, spokesmen for the German military establishment exaggerated the threat to Germany and the benefits of empire for organizationally self-serving reasons. Indeed, members of the German elite sometimes privately acknowledged that Germany was under less threat than the public was being told. For example, the Secretary of State in the Foreign Office, Kiderlen-Wächter, admitted, "If we do not conjure up a war into being, no one else certainly will do so," since "The Republican government of France is certainly peace-minded. The British do not want war. They will never give cause for it. . . ."[46]

Nevertheless, the German public believed that German security was precarious, and security arguments formed the core of the public case for expansion. Moreover, these arguments proved persuasive, and the chauvinist public climate which they created enabled the elite to pursue expansion, whatever elite motivation might actually have been. Indeed, some members of the German government eventually felt pushed into reckless action by an extreme chauvinist public opinion which they felt powerless to resist. Admiral von Müller later explained that Germany pursued a bellicose policy during the July crisis because "The government, already weakened by domestic disunity, found itself inevitably under pressure from a great part of the German people which had been whipped into a high-grade chauvinism by Navalists and Pan-Germans."[47] Bethmann Hollweg felt his hands tied by an expansionist public climate: "With these idiots [the Pan-Germans] one cannot conduct a foreign policy—on the contrary. Together with other factors they will eventually make any reasonable course impossible for us."[48] Thus the search for security was a fundamental cause of German conduct, whether or not the elite was motivated by security concerns, because the elite was

---

45. Examples are: Arno Mayer, "Domestic Causes of the First World War," in Leonard Krieger and Fritz Stern, eds., *The Responsibility of Power* (New York: Macmillan, 1968), pp. 286–300; Berghahn, *Germany and the Approach of War;* Fischer, *War of Illusions,* pp. 257–258; and Imanuel Geiss, *German Foreign Policy, 1871–1914* (Boston: Routledge & Kegan Paul, 1976). A criticism is Marc Trachtenberg, "The Social Interpretation of Foreign Policy," *Review of Politics,* Vol. 40, No. 3 (July 1978), pp. 341–350.

46. In 1910, quoted in Geiss, *German Foreign Policy,* p. 126.

47. Admiral von Müller, quoted in Fritz Stern, *The Failure of Illiberalism* (London: Allen & Unwin, 1972), p. 94.

48. In 1909, quoted in Konrad H. Jarausch, *The Enigmatic Chancellor: Bethmann Hollweg and the Hubris of Imperial Germany* (New Haven: Yale University Press, 1973), p. 119. See also ibid., p. 152; and Geiss, *German Foreign Policy,* pp. 135–137. As Jules Cambon, French ambassador to Germany, perceptively remarked: "It is false that in Germany the nation is peaceful and the government bellicose—the exact opposite is true." In 1911, quoted in Jarausch, *Enigmatic Chancellor,* p. 125.

allowed or even compelled to adopt expansionist policies by a German public which found security arguments persuasive.

The same mixture of insecurity and perceived opportunity stiffened resistance to German expansion and fuelled a milder expansionism elsewhere in Europe, intensifying the conflict between Germany and its neighbors. In France the nationalist revival and French endorsement of a firm Russian policy in the Balkans were inspired partly by a growing fear of the German threat after 1911,[49] partly by an associated concern that Austrian expansion in the Balkans could shift the European balance of power in favor of the Central Powers and thereby threaten French security, and partly by belief that a war could create opportunities for French expansion. The stiffer French "new attitude" on Balkan questions in 1912 was ascribed to the French belief that "a territorial acquisition on the part of Austria would affect the general balance of power in Europe and as a result touch the particular interests of France"—a belief which assumed that the power balance was relatively precarious, which in turn assumed a world of relatively strong offense.[50] At the same time some Frenchmen looked forward to "a beautiful war which will deliver all the captives of Germanism,"[51] inspired by a faith in the power of the offensive that was typified by the enthusiasm of Joffre's deputy, General de Castelnau: "Give me 700,000 men and I will conquer Europe!"[52]

Russian policy in the Balkans was driven both by fear that Austrian expansion could threaten Russian security and by hopes that Russia could destroy its enemies if war developed under the right conditions. Sazonov saw a German–Austrian Balkan program to "deliver the Slavonic East, bound hand and foot, into the power of Austria–Hungary," followed by the German seizure of Constantinople, which would gravely threaten Russian security by placing all of Southern Russia at the mercy of German power.[53] Eventually a "German Khalifate" would be established, "extending from the banks of the Rhine to the mouth of the Tigris and Euphrates," which would reduce

49. See Eugen Weber, *The Nationalist Revival in France, 1905–1914* (Berkeley and Los Angeles: University of California Press, 1968), passim; and Snyder, "Defending the Offensive," pp. 32–33.
50. By the Russian ambassador to Paris, A.P. Izvolsky, quoted in Schmitt, *Coming of the War,* Vol. 1, p. 21.
51. *La France Militaire,* in 1913, quoted in Weber, *Nationalist Revival in France,* p. 127.
52. In 1913, quoted in L.C.F. Turner, *Origins of the First World War* (London: Edward Arnold, 1970), p. 53.
53. Serge Sazonov, *Fateful Years, 1909–1916* (London: Jonathan Cape, 1928), p. 179. See also Schmitt, *Coming of the War,* Vol. 1, p. 87.

"Russia to a pitiful dependence upon the arbitrary will of the Central Powers."[54] At the same time some Russians believed these threats could be addressed by offensive action: Russian leaders spoke of the day when "the moment for the downfall of Austria–Hungary arrives,"[55] and the occasion when "The Austro-Hungarian ulcer, which today is not yet so ripe as the Turkish, may be cut up."[56] Russian military officers contended that "the Austrian army represents a serious force. . . . But on the occasion of the first great defeats all of this multi-national and artificially united mass ought to disintegrate."[57]

In short, the belief that conquest was easy and security scarce was an important source of German–Entente conflict. Without it, both sides could have adopted less aggressive and more accommodative policies.

THE INCENTIVE TO PREEMPT

American strategists have long assumed that World War I was a preemptive war, but they have not clarified whether or how this was true.[58] Hence two questions should be resolved to assess the consequences of the cult of the offensive: did the states of Europe perceive an incentive to move first in 1914, which helped spur them to mobilize or attack? If so, did the cult of the offensive help to give rise to this perception?

The question of whether the war was preemptive reduces to the question of why five principal actions in the July crisis were taken. These actions are: the Russian preliminary mobilization ordered on July 25–26; the partial Russian mobilization against Austria–Hungary ordered on July 29; the Russian

---

54. Sazonov, *Fateful Years,* pp. 191, 204.
55. Izvolsky, in 1909, quoted in Schmitt, *Coming of the War,* Vol. 1, p. 129.
56. Sazonov, in 1913, quoted in ibid., p. 135.
57. *Sbornik glavnogo upravleniia general'nogo shtaba,* the secret magazine of the Russian general staff, in 1913, quoted in William C. Fuller, "The Russian Empire and Its Potential Enemies" (manuscript, 1980), p. 21.

British resistance was also driven by security concerns: during the July crisis the London *Times* warned that "the ruin of France or the Low Countries would be the prelude to our own," while other interventionists warned that Antwerp in German hands would be a "pistol pointed at the heart of England," and that the German threat to France and the Low Countries created "a deadly peril for ourselves." The *Times* on August 4, quoted in Geoffrey Marcus, *Before the Lamps Went Out* (Boston: Little, Brown, 1965), p. 305; and the *Pall Mall Gazette* and James Gavin, on July 29 and August 2, quoted in ibid., pp. 243, 268.

58. Suggesting that World War I was preemptive are: Herman Kahn, *On Thermonuclear War,* 2nd ed. (New York: The Free Press, 1969), pp. 359–362; Schelling, *Arms and Influence,* pp. 223–224; Jervis, "Cooperation under the Security Dilemma," pp. 191–192; Quester, *Offense and Defense,* pp. 110–111; Richard Ned Lebow, *Between Peace and War: The Nature of International Crisis* (Baltimore, Md.: The Johns Hopkins University Press, 1981), pp. 238–242.

full mobilization ordered on July 30; French preliminary mobilization measures ordered during July 25–30; and the German attack on the Belgian fortress at Liège at the beginning of the war. The war was preemptive if Russia and France mobilized preemptively, since these mobilizations spurred German and Austrian mobilization, opening windows which helped cause war. Thus while the mobilizations were not acts of war, they caused effects which caused war. The war was also preemptive if Germany struck Liège preemptively, since the imperative to strike Liège was one reason why "mobilization meant war" to Germany.

The motives for these acts cannot be determined with finality; testimony by the actors is spotty and other direct evidence is scarce. Instead, motives must be surmised from preexisting beliefs, deduced from circumstances, and inferred from clues which may by themselves be inconclusive. However, three pieces of evidence suggest that important preemptive incentives existed, and helped to shape conduct. First, most European leaders apparently believed that mobilization by either side which was not answered within a very few days, or even hours, could affect the outcome of the war. This judgment is reflected both in the length of time which officials assumed would constitute a militarily significant delay between mobilization and offsetting counter-mobilization, and in the severity of the consequences which they assumed would follow if they mobilized later than their opponents.

Second, many officials apparently assumed that significant mobilization measures and preparations to attack could be kept secret for a brief but significant period. Since most officials also believed that a brief unanswered mobilization could be decisive, they concluded that the side which mobilized first would have the upper hand.

Third, governments carried out some of their mobilization measures in secrecy, suggesting that they believed secret measures were feasible and worthwhile.

THE PERCEIVED SIGNIFICANCE OF SHORT DELAYS. Before and during the July crisis European leaders used language suggesting that they believed a lead in ordering mobilization of roughly one to three days would be significant. In Austria, General Conrad believed that "every day was of far-reaching importance," since "any delay might leave the [Austrian] forces now assembling in Galicia open to being struck by the full weight of a Russian offensive in the midst of their deployment."[59] In France, Marshall Joffre warned the

59. July 29, quoted in Albertini, *Origins*, Vol. 2, p. 670.

French cabinet that "any delay of twenty-four hours in calling up our reservists" once German preparations began would cost France "ten to twelve miles for each day of delay; in other words, the initial abandonment of much of our territory."[60] In Britain, one official believed that France "cannot possibly delay her own mobilization for even the fraction of a day" once Germany began to mobilize.[61]

In Germany, one analyst wrote that "A delay of a single day . . . can scarcely ever be rectified."[62] Likewise Moltke, on receiving reports of preparations in France and Russia during the July crisis, warned that "the military situation is becoming from day to day more unfavorable for us," and would "lead to fateful consequences for us" if Germany did not respond.[63] On July 30 he encouraged Austria to mobilize, warning that "every hour of delay makes the situation worse, for Russia gains a start."[64] On August 1, the Prussian ministry of war was reportedly "very indignant over the day lost for the mobilization" by the German failure to mobilize on July 30.[65] The German press drove home the point that if mobilization by the adversary went unanswered even briefly, the result could be fatal, one German newspaper warning that "Every delay [in mobilizing] would cost us an endless amount of blood" if Germany's enemies gained the initiative; hence "it would be disastrous if we let ourselves be moved by words not to carry on our preparations so quickly. . . ."[66]

---

60. July 29, from Marshall Joffre, *The Personal Memoirs of Marshall Joffre*, 2 vols., trans. T. Bentley Mott (New York: Harper & Brothers, 1932), Vol. 1, p. 125.
61. Eyre Crowe, on July 27, quoted in Geiss, *July 1914*, p. 251.
62. Kraft zu Hohenlohe-Ingelfingen, in 1898, quoted in Ropp, *War in the Modern World*, p. 203.
63. To Bethmann Hollweg, on July 29, quoted in Geiss, *July 1914*, p. 284.
64. Quoted in Schmitt, *Coming of the War*, Vol. 2, p. 196.
65. Ibid., p. 265n.
66. The *Reinisch-Westfälische Zeitung*, July 31, quoted in Jonathan French Scott, *The Five Weeks* (New York: John Day Co., 1927), p. 146.

Likewise after the war General von Kluck, who commanded the right wing of the German army in the march on Paris, claimed that if the German army had been mobilized and deployed "three days earlier, a more sweeping victory and decisive result would probably have been gained" against France, and Admiral Tirpitz complained that German diplomats had given Britain and Belgium several crucial days warning of the German attack on July 29, which "had an extraordinarily unfavorable influence on the whole course of the war." A delay of "only a few days" in the preparation of the British expeditionary force "might have been of the greatest importance to us." Schmitt, *Coming of the War*, Vol. 2, p. 148n.; and Albertini, *Origins*, Vol. 3, p. 242n.

A more relaxed opinion was expressed by the Prussian war minister, General Falkenhayn, who seemed to feel that it would be acceptable if German mobilization "follows two or three days later than the Russian and Austrian," since it "will still be completed more quickly than theirs." Schmitt, *Coming of the War*, Vol. 2, p. 147. However, he also expressed himself in favor

Thus time was measured in small units: "three days," "day to day," "a single day," "the fraction of a day," or even "every hour." Moreover, the consequences of conceding the initiative to the adversary were thought to be extreme. The Russian Minister of Agriculture, Alexander Krivoshein, warned that if Russia delayed its mobilization "we should be marching toward a certain catastrophe,"[67] and General Janushkevich warned the Russian foreign minister that "we were in danger of losing [the war] before we had time to unsheath our sword" by failing to mobilize promptly against Germany.[68] General Joffre feared that France would find itself "in an irreparable state of inferiority" if it were outstripped by German mobilization.[69] And in Germany, officials foresaw dire consequences if Germany conceded the initiative either in the East or the West. Bethmann Hollweg explained to one of his ambassadors that if German mobilization failed to keep pace with the Russian, Germany would suffer large territorial losses: "East Prussia, West Prussia, and perhaps also Posen and Silesia [would be] at the mercy of the Russians."[70] Such inaction would be "a crime against the safety of our fatherland."[71]

Germans also placed a high value on gaining the initiative at Liège, since Liège controlled a vital Belgian railroad junction, and German forces could not seize Liège with its tunnels and bridges intact unless they surprised the Belgians. As Moltke wrote before the war, the advance through Belgium "will hardly be possible unless Liège is in our hands . . . the possession of Liège is the *sine qua non* of our advance." But seizing Liège would require "meticulous preparation and surprise" and "is only possible if the attack is made at once, before the areas between the forts are fortified," "immediately" after the declaration of war.[72] In short, the entire German war plan would be ruined if Germany allowed Belgium to prepare the defense of Liège.

This belief that brief unanswered preparations and actions could be decisive reflected the implicit assumption that the offense had the advantage. Late mobilization would cost Germany control of East and West Prussia only

---

of preemption at other junctures. See ibid., p. 297; and Berghahn, *Germany and the Approach of War*, p. 203.
67. To Sazonov, July 30, quoted in Geiss, *July 1914*, p. 311.
68. To Sazonov, July 30, quoted in Albertini, *Origins*, Vol. 2, p. 566.
69. August 1, Poincaré reporting Joffre's view, quoted in Albertini, *Origins*, Vol. 3, p. 100.
70. August 1, quoted in Schmitt, *Coming of the War*, Vol. 2, p. 264.
71. August 1, quoted in Albertini, *Origins*, Vol. 3, p. 167.
72. Ritter, *The Schlieffen Plan*, p. 166. On the Liège attack, see also Snyder, "Defending the Offensive," pp. 203, 285–287.

if Russian offensive power were strong, and German defensive power were weak; mobilizing late could only be a "crime against the safety" of Germany if numerically superior enemies could destroy it; lateness could only confront Russia with "certain catastrophe" or leave it in danger of "losing before we have time to unsheath our sword" if Germany could develop a powerful offensive with the material advantage it would gain by preparing first; and lateness could only condemn France to "irreparable inferiority" if small material inferiority translated into large territorial losses. Had statesmen understood that in reality the defense had the advantage, they also would have known that the possession of the initiative could not be decisive, and could have conceded it more easily.

WAS SECRET PREPARATION BELIEVED FEASIBLE? The belief that delay could be fatal would have created no impulse to go first had European leaders believed that they could detect and offset their opponents' preparations immediately. However, many officials believed that secret action for a short time was possible. Russian officials apparently lacked confidence in their own ability to detect German or Austrian mobilization, and their decisions to mobilize seem to have been motivated partly by the desire to forestall surprise preparation by their adversaries. Sazonov reportedly requested full mobilization on July 30 partly from fear that otherwise Germany would "gain time to complete her preparations in secret."[73] Sazonov offers confirmation in his memoirs, explaining that he had advised mobilization believing that "The perfection of the German military organization made it possible by means of personal notices to the reservists to accomplish a great part of the work quietly." Germany could then "complete the mobilization in a very short time. This circumstance gave a tremendous advantage to Germany, but we could counteract it to a certain extent by taking measures for our own mobilization in good time."[74]

Similar reasoning contributed to the Russian decision to mobilize against Austria on July 29. Sazonov explains that the mobilization was undertaken in part "so as to avoid the danger of being taken unawares by the Austrian

---

73. Paleologue's diary, quoted in Albertini, *Origins,* Vol. 2, p. 619.
74. Sazonov, *Fateful Years,* pp. 202–203. The memorandum of the day of the Russian foreign ministry for July 29 records that Russian officials had considered whether Germany seriously sought peace, or whether its diplomacy "was only intended to lull us to sleep and so to postpone the Russian mobilization and thus gain time wherein to make corresponding preparations." Quoted in Geiss, *July 1914,* pp. 296–297.

preparations."[75] Moreover, recent experience had fuelled Russian fears of an Austrian surprise: during the Balkan crisis of 1912, the Russian army had been horrified to discover that Austria had secretly mobilized in Galicia, without detection by Russian intelligence; and this experience resolved the Russian command not to be caught napping again. In one observer's opinion, "the experience of 1912 . . . was not without influence as regards Russia's unwillingness to put off her mobilization in the July days of 1914."[76]

Top Russian officials also apparently believed that Russia could itself mobilize secretly, and some historians ascribe the Russian decision to mobilize partly to this erroneous belief. Luigi Albertini writes that Sazonov did not realize that the mobilization order would be posted publicly and that, accordingly, he "thought Russia could mobilize without Germany's knowing of it immediately."[77] Albertini reports that the German ambassador caused "real stupefaction" by appearing at the Russian ministry for foreign affairs with a red mobilization poster on the morning of mobilization,[78] and concludes that the "belief that it was possible to proceed to general mobilization without making it public may well have made Sazonov more inclined to order it."[79]

Contemporary accounts confirm that the Russian leadership believed in their own ability to mobilize in secret. The memorandum of the Russian Ministry for Foreign Affairs records that Sazonov sought to "proceed to the general mobilization as far as possible secretly and without making any public announcement concerning it," in order "To avoid rendering more acute our relations with Germany."[80] And in informing his government of Russian preliminary mobilization measures which began on July 26, the French ambassador indicated Russian hopes that they could maintain secrecy: "Secret preparations will, however, commence already today,"[81] and "the military districts of Warsaw, Vilna and St. Petersburg are secretly making preparations."[82] His telegram informing Paris of Russian general mobilization ex-

75. Sazonov, *Fateful Years*, p. 188.
76. A.M. Zayonchovsky, quoted in Lieven, *Russia and the Origins of the First World War*, p. 149.
77. Albertini, *Origins*, Vol. 2, p. 624.
78. Ibid., quoting Taube who quoted Nolde.
79. Ibid., p. 573. See also p. 584, suggesting that "Sazonov was such a greenhorn in military matters as to imagine the thing could be done, and was only convinced of the contrary when on 31 July he saw the red notices, calling up reservists, posted up in the streets of St. Petersburg." This point "provides the key to many mysteries" (p. 624).
80. For July 31, in Geiss, *July 1914*, p. 326.
81. Paleologue, July 25, in Albertini, *Origins*, Vol. 2, p. 591.
82. Paleologue, July 26, in ibid., p. 592.

plained that "the Russian government has decided to proceed secretly to the first measures of general mobilization."[83]

Like their Russian counterparts, top French officials also apparently feared that Germany might mobilize in secret, which spurred the French to their own measures. Thus during the July crisis General Joffre spoke of "the concealments [of mobilization] which are possible in Germany,"[84] and referred to "information from excellent sources [which] led us to fear that on the Russian front a sort of secret mobilization was taking place [in Germany]."[85] In his memoirs, Joffre quotes a German military planning document acquired by the French government before the July crisis, which he apparently took to indicate German capabilities, and which suggested that Germany could take "quiet measures . . . in preparation for mobilization," including "a discreet assembly of complementary personnel and materiel" which would "assure us advantages very difficult for other armies to realize in the same degree."[86] The French ambassador to Berlin, Jules Cambon, also apparently believed that Germany could conduct preliminary mobilization measures in secret, became persuaded during the July crisis that it had in fact done this, and so informed Paris: "In view of German habits, [preliminary measures] can be taken without exciting the population or causing indiscretions to be committed. . . ."[87] For their part the Germans apparently did not believe that they or their enemies could mobilize secretly, but they did speak in terms suggesting that Germany could surprise the Belgians: German planners referred to the "*coup de main*" at Liège and the need for "meticulous preparation and surprise."[88]

To sum up, then, French policymakers feared that Germany could mobilize secretly; Russians feared secret mobilization by Germany or Austria, and hoped Russian mobilization could be secret; while Central Powers planners

---

83. Ibid., p. 620.
84. August 1, quoted in Joffre, *Personal Memoirs*, p. 128.
85. July 29, quoted in ibid., p. 120.
86. Ibid., p. 127.
87. Cambon dispatch to Paris, July 21, quoted in ibid., p. 119. Joffre records that Cambon's telegram, which mysteriously did not arrive in Paris until July 28, convinced him that "for seven days at least the Germans had been putting into effect the plan devised for periods of political tension and that our normal methods of investigation had not revealed this fact to us. Our adversaries could thus reach a condition of mobilization that was almost complete," reflecting Joffre's assumption that secret German measures were possible.
88. Moltke, quoted in Ritter, *The Schlieffen Plan*, p. 166.

saw less possibility for preemptive mobilization by either side, but hoped to mount a surprise attack on Belgium.[89]

DID STATESMEN ACT SECRETLY? During the July crisis European statesmen sometimes informed their opponents before they took military measures, but on other occasions they acted secretly, suggesting that they believed the initiative was both attainable and worth attaining, and indicating that the desire to seize the initiative may have entered into their decisions to mobilize. German leaders warned the French of their preliminary measures taken on July 29,[90] and their pre-mobilization and mobilization measures taken on July 31;[91] and they openly warned the Russians on July 29 that they would mobilize if Russia conducted a partial mobilization.[92] Russia openly warned Austria on July 27 that it would mobilize if Austria crossed the Serbian frontier,[93] and then on July 28 and July 29 openly announced to Germany and Austria its partial mobilization of July 29,[94] and France delayed full mobilization until after Germany had taken the onus on itself by issuing ultimata to Russia and France. However, Russia, France, and Germany tried

89. During the July crisis, adversaries actually detected signs of most major secret mobilization activity in roughly 6–18 hours, and took responsive decisions in 1–2 days. Accordingly, the maximum "first mobilization advantage" which a state could gain by forestalling an adversary who otherwise would have begun mobilizing first was roughly 2–4 days. Orders for Russian preliminary mobilization measures were issued in sequential telegrams transmitted between 4:00 p.m. on July 25 and 3:26 a.m. on July 26; Berlin received its first reports of these measures early on July 26; and at 4:00 p.m. on July 27 the German intelligence board concluded that Russian premobilization had in fact begun, for a lag of roughly one and one-half to two days between the issuance of orders and their definite detection. Sidney B. Fay, *The Origins of the World War*, 2 vols., 2nd ed. rev. (New York: Free Press, 1966), Vol. 2, pp. 310–315; and Ulrich Trumpener, "War Premeditated? German Intelligence Operations in July 1914," *Central European History*, Vol. 9 (1976), pp. 67–70. Full Russian mobilization was ordered at 6:00 p.m. on July 30, first rumors reached Berlin very late on July 30, more definite but inconclusive information was received around 7:00 a.m. July 31, reliable confirmation was received at 11:45 a.m., and German preliminary mobilization was ordered at 1:00 p.m., for a lag of roughly 20 hours. Fay, *Origins of the World War*, Vol. 2, p. 473; Schmitt, *Coming of the War*, Vol. 2, pp. 211–212, 262–265; and Trumpener, "War Premeditated?," pp. 80–83. French preliminary measures were begun on July 25, expanded on July 26, further expanded on July 27, and remained substantially undetected on July 28. Secondary sources do not clarify when Germany detected French preliminary measures, but it seems that German discovery lagged roughly two days behind French actions. Schmitt, *Coming of the War*, Vol. 2, pp. 17–19; Joffre, *Personal Memoirs*, pp. 115–118; and Trumpener, "War Premeditated?," pp. 71–73. As for Liège, it was not captured as quickly as German planners had hoped, but was not properly defended when the Germans arrived, and was taken in time to allow the advance into France.

90. Albertini, *Origins*, Vol. 2, p. 491.

91. Schmitt, *Coming of the War*, Vol. 2, pp. 267–268.

92. Ibid., p. 105.

93. Albertini, *Origins*, Vol. 2, p. 529.

94. Ibid., pp. 549, 551; and Geiss, *July 1914*, pp. 262, 278, 299.

to conceal four of the five major preemptive actions of the crisis: the Russians hid both their preliminary measures of July 25–26 and their general mobilization of July 30, the French attempted to conceal their preliminary mobilization measures of July 25–29, and the Germans took great care to conceal their planned *coup de main* against Liège. Thus states sometimes conceded the initiative, but sought it at critical junctures.

Overall, evidence suggests that European leaders saw some advantage to moving first in 1914: the lags which they believed significant lay in the same range as the lags they believed they could gain or forestall by mobilizing first. These perceptions probably helped spur French and Russian decisions to mobilize, which in turn helped set in train the German mobilization, which in turn meant war partly because the Germans were determined to preempt Liège. Hence the war was in some modest measure preemptive.

If so, the cult of the offensive bears some responsibility. Without it, statesmen would not have thought that secret mobilization or preemptive attack could be decisive. The cult was not the sole cause of the perceived incentive to preempt; rather, three causes acted together, the others being the belief that mobilization could briefly be conducted secretly, and the systems of reserve manpower mobilization which enabled armies to multiply their strength in two weeks. The cult had its effect by magnifying the importance of these other factors in the minds of statesmen, which magnified the incentive to preempt which these factors caused them to perceive. The danger that Germany might gain time to complete preparations in secret could only alarm France and Russia if Germany could follow up these preparations with an effective offensive; otherwise, early secret mobilization could *not* give "a tremendous advantage" to Germany, and such a prospect would not require a forestalling response. Sazonov could have been tempted to mobilize secretly only if early Russian mobilization would forestall important German gains, or could provide important gains for Russia, as could only have happened if the offense were powerful.

"WINDOWS" AND PREVENTIVE WAR

Germany and Austria pursued bellicose policies in 1914 partly to shut the looming "windows" of vulnerability which they envisioned lying ahead, and partly to exploit the brief window of opportunity which they thought the summer crisis opened. This window logic, in turn, grew partly from the cult of the offensive, since it depended upon the implicit assumption that the offense was strong. The shifts in the relative sizes of armies, economies, and alliances which fascinated and frightened statesmen in 1914 could have cast such a long shadow only in a world where material advantage promised decisive results in warfare, as it could only in an offense-dominant world.

The official communications of German leaders are filled with warnings that German power was in relative decline, and that Germany was doomed unless it took drastic action—such as provoking and winning a great crisis which would shatter the Entente, or directly instigating a "great liquidation" (as one general put it).[95] German officials repeatedly warned that Russian military power would expand rapidly between 1914 and 1917, as Russia carried out its 1913–1914 Great Program, and that in the long run Russian power would further outstrip German power because Russian resources were greater.[96] In German eyes this threat forced Germany to act. Secretary of State Jagow summarized a view common in Germany in a telegram to one of his ambassadors just before the July crisis broke:

> Russia will be ready to fight in a few years. Then she will crush us by the number of her soldiers; then she will have built her Baltic fleet and her strategic railways. Our group in the meantime will have become steadily weaker. . . . I do not desire a preventive war, but if the conflict should offer itself, we ought not to shirk it.[97]

Similarly, shortly before Sarajevo the Kaiser reportedly believed that "the big Russian railway constructions were . . . preparations for a great war which could start in 1916" and wondered "whether it might not be better to attack than to wait."[98] At about the same time Chancellor Bethmann Hollweg declared bleakly, "The future belongs to Russia which grows and grows and becomes an even greater nightmare to us,"[99] warning that "After the completion of their strategic railroads in Poland our position [will be] untenable."[100] During the war, Bethmann confessed that the "window" argument

95. Von Plessen, quoted in Isabell V. Hull, *The Entourage of Kaiser Wilhelm II, 1888–1918* (New York: Cambridge University Press, 1982), p. 261. Thus Bethmann summarized German thinking when he suggested on July 8 that the Sarajevo assassination provided an opportunity either for a war which "we have the prospect of winning" or a crisis in which "we still certainly have the prospect of maneuvering the Entente apart. . . ." Thompson, *In the Eye of the Storm*, p. 75.
96. The Russian program planned a 40 percent increase in the size of the peacetime Russian army and a 29 percent increase in the number of officers over four years. Lieven, *Russia & the Origins of the First World War*, p. 111.
97. July 18, quoted in Schmitt, *Coming of the War*, Vol. 1, p. 321.
98. June 21, quoted in Fischer, *War of Illusions*, p. 471, quoting Max Warburg.
99. July 7, quoted in ibid., p. 224, quoting Riezler.
100. July 7, quoted in Jarausch, "The Illusion of Limited War," p. 57. Likewise on July 20, he expressed terror at Russia's "growing demands and colossal explosive power. In a few years

had driven German policy in 1914: "Lord yes, in a certain sense it was a preventive war," motivated by "the constant threat of attack, the greater likelihood of its inevitability in the future, and by the military's claim: today war is still possible without defeat, but not in two years!"[101]

Window logic was especially prevalent among the German military officers, many of whom openly argued for preventive war during the years before the July crisis. General Moltke declared, "I believe a war to be unavoidable and: the sooner the better" at the infamous "war council" of December 8, 1912,[102] and he expressed similar views to his Austrian counterpart, General Conrad, in May 1914: "to wait any longer meant a diminishing of our chances; as far as manpower is concerned, one cannot enter into a competition with Russia,"[103] and "We [the German Army] are ready, the sooner the better for us."[104] During the July crisis Moltke remarked that "we shall never hit it again so well as we do now with France's and Russia's expansion of their armies incomplete," and argued that "the singularly favorable situation be exploited for military action."[105] After the war Jagow recalled a conversation with Moltke in May 1914, in which Moltke had spelled out his reasoning:

> In two–three years Russia would have completed her armaments. The military superiority of our enemies would then be so great that he did not know how we could overcome them. Today we would still be a match for them. In his opinion there was no alternative to making preventive war in order to defeat the enemy while we still had a chance of victory. The Chief of General Staff therefore proposed that I should conduct a policy with the aim of provoking a war in the near future.[106]

Other members of the German military shared Moltke's views, pressing for preventive war because "conditions and prospects would never become

---

she would be supreme—and Germany her first lonely victim." Quoted in Lebow, *Between Peace and War*, p. 258n.

101. Jarausch, "The Illusion of Limited War," p. 48. Likewise Friedrich Thimme quoted Bethmann during the war: "He also admits that our military are quite convinced that they could still be victorious in the war, but that in a few years time, say in 1916 after the completion of Russia's railway network, they could not. This, of course, also affected the way in which the Serbian question was dealt with." Quoted in Volker R. Berghahn and Martin Kitchen, eds., *Germany in the Age of Total War* (Totowa, N.J.: Barnes and Noble, 1981), p. 45.

102. Fischer, *War of Illusions*, p. 162.

103. Berghahn, *Germany and the Approach of War*, p. 171.

104. Geiss, *German Foreign Policy*, p. 149.

105. Berghahn, *Germany and the Approach of War*, p. 203.

106. Quoted in J.C.G. Röhl, ed., *From Bismarck to Hitler: The Problem of Continuity in German History* (London: Longman, 1970), p. 70.

better."[107] General Gebstattel recorded the mood of the German leadership on the eve of the war: "Chances better than in two or three years hence and the General Staff is reported to be confidently awaiting events."[108] The Berlin *Post,* a newspaper which often reflected the views of the General Staff, saw a window in 1914: "at the moment the state of things is favorable for us. France is not yet ready for war. England has internal and colonial difficulties, and Russia recoils from the conflict because she fears revolution at home. Ought we to wait until our adversaries are ready?" It concluded that Germany should "prepare for the inevitable war with energy and foresight" and "begin it under the most favorable conditions."[109]

German leaders also saw a tactical window of opportunity in the political constellation of July 1914, encouraging them to shut their strategic window of vulnerability. In German eyes, the Sarajevo assassination created favorable conditions for a confrontation, since it guaranteed that Austria would join Germany against Russia and France (as it might not if war broke out over a colonial conflict or a dispute in Western Europe), and it provided the Central powers with a plausible excuse, which raised hopes that Britain might remain neutral. On July 8, Bethmann Hollweg reportedly remarked, "If war comes from the east so that we have to fight for Austria–Hungary and not Austria–Hungary for us, we have a chance of winning."[110] Likewise, the German ambassador to Rome reportedly believed on July 27 that "the present moment is extraordinarily favorable to Germany,"[111] and the German ambassador to London even warned the British Prime Minister that "there was some feeling in Germany . . . that trouble was bound to come and therefore it would be better not to restrain Austria and let trouble come now, rather than later."[112]

The window logic reflected in these statements is a key to German conduct in 1914: whether the Germans were aggressive or restrained depended on

107. Leuckart's summary of the views of the General Staff, quoted in Geiss, *July 1914,* p. 69. For more on advocacy of preventive war by the German army, see Martin Kitchen, *The German Officer Corps, 1890–1914* (Oxford: Clarendon Press, 1968), pp. 96–114; and Hull, *Entourage of Kaiser Wilhelm II,* pp. 236–265.
108. August 2, quoted in Fischer, *War of Illusions,* p. 403.
109. February 24, 1914, in Schmitt, *Coming of the War,* Vol. 1, p. 100n.; and Fischer, *War of Illusions,* pp. 371–272.
110. Jarausch, "Illusion of Limited War," p. 58. Earlier Bülow had explained why the Agadir crisis was an unsuitable occasion for war in similar terms: "In 1911 the situation was much worse. The complication would have begun with Britain; France would have remained passive, it would have forced us to attack and then there would have been no *causus foederis* for Austria . . . whereas Russia was obliged to join in." In 1912, quoted in Fischer, *War of Illusions,* p. 85.
111. Schmitt, *Coming of the War,* Vol. 2, p. 66n.
112. Ibid., Vol. 1, p. 324, quoting Lichnowsky, on July 6.

whether at a given moment they thought windows were open or closed. Germany courted war on the Balkan question after Sarajevo because window logic led German leaders to conclude that war could not be much worse than peace, and might even be better, if Germany could provoke the right war under the right conditions against the right opponents. German leaders probably preferred the status quo to a world war against the entire Entente, but evidence suggests that they also preferred a continental war against France and Russia to the status quo—as long as Austria joined the war, and as long as they could also find a suitable pretext which they could use to persuade the German public that Germany fought for a just cause. This, in turn, required that Germany engineer a war which engaged Austrian interests, and in which Germany could cast itself as the attacked, in order to involve the Austrian army, to persuade Britain to remain neutral, and to win German public support. These window considerations help explain both the German decision to force the Balkan crisis to a head and German efforts to defuse the crisis after it realized that it had failed to gain British neutrality. The German peace efforts after July 29 probably represent a belated effort to reverse course after it became clear that the July crisis was not such an opportune war window after all.

Window logic also helped to persuade Austria to play the provocateur for Germany. Like their German counterparts, many Austrian officials believed that the relative strength of the central powers was declining, and saw in Sarajevo a rare opportunity to halt this decline by force. Thus the Austrian War Minister, General Krobatin, argued in early July that "it would be better to go to war immediately, rather than at some later period, because the balance of power must in the course of time change to our disadvantage," while the Austrian Foreign Minister, Count Berchtold, favored action because "our situation must become more precarious as time goes on,"[113] warning that unless Austria destroyed the Serbian army in 1914, it would face "another attack [by] Serbia in much more unfavorable conditions" in two or three years.[114] Likewise, the Austrian foreign ministry reportedly believed that, "if Russia would not permit the localization of the conflict with Serbia, the present moment was more favorable for a reckoning than a later one would be";[115] General Conrad believed, "If it comes to war with Russia—as

113. July 7, quoted in Geiss, *July 1914*, pp. 81, 84.
114. July 31, quoted in Schmitt, *Coming of the War*, Vol. 2, p. 218.
115. Ibid., Vol. 1, p. 372, quoting Baron von Tucher on July 18.

it must some day—today is as good as any other day";[116] and the Austrian ambassador to Italy believed an Austro–Serbian war would be "a piece of real good fortune," since "for the Triple Alliance the present moment is more favorable than another later."[117]

Thus the First World War was in part a "preventive" war, launched by the Central powers in the belief that they were saving themselves from a worse fate in later years. The cult of the offensive bears some responsibility for that belief, for in a defense-dominated world the windows which underlie the logic of preventive war are shrunken in size, as the balance of power grows less elastic to the relative sizes of armies and economies; and windows cannot be shut as easily by military action. Only in a world taken by the cult of the offensive could the window logic which governed German and Austrian conduct have proved so persuasive: Germans could only have feared that an unchecked Russia could eventually "crush us by the numbers of her soldiers," or have seen a "singularly favorable situation" in 1914 which could be "exploited by military action" if material superiority would endow the German and Russian armies with the ability to conduct decisive offensive operations against one another. Moltke claimed he saw "no alternative to making preventive war," but had he believed that the defense dominated, better alternatives would have been obvious.

The cult of the offensive also helped cause the arms race before 1914 which engendered the uneven rates of military growth that gave rise to visions of windows. The German army buildup after 1912 was justified by security arguments: Bethmann Hollweg proclaimed, "For Germany, in the heart of Europe, with open boundaries on all sides, a strong army is the most secure guarantee of peace," while the Kaiser wrote that Germany needed "More ships and soldiers . . . because our existence is at stake."[118] This buildup provoked an even larger Russian and French buildup, which created the windows which alarmed Germany in 1914.[119] Thus the cult both magnified

116. In October 1913, quoted in Gerhard Ritter, *The Sword and the Scepter: The Problem of Militarism in Germany*, 4 vols., trans. Heinz Norden (Coral Gables, Fla.: University of Miami Press, 1969–73), Vol. 2, p. 234. Likewise the *Militärisch Rundschau* argued for provoking war: "Since we shall have to accept the contest some day, let us provoke it at once." On July 15, 1914, quoted in Schmitt, *Coming of the War*, Vol. 1, p. 367. For more on preventive war and the Austrian army, see Ritter, *Sword and the Scepter*, Vol. 2, pp. 227–239.

117. Count Merey, July 29, quoted in Albertini, *Origins*, Vol. 2, p. 383.

118. Both in 1912, quoted in Jarausch, *Enigmatic Chancellor*, p. 95; and Fischer, *War of Illusions*, p. 165.

119. On the motives for the Russian buildup, see P.A. Zhilin, "Bol'shaia programma po usileniiu russkoi armii," *Voenno-istoricheskii zhurnal*, No. 7 (July 1974), pp. 90–97.

the importance of fluctuations in ratios of forces and helped to fuel the arms race which fostered them.

THE SCOPE AND INFLEXIBILITY OF MOBILIZATION PLANS

The spreading of World War I outward from the Balkans is often ascribed to the scope and rigidity of the Russian and German plans for mobilization, which required that Russia must also mobilize armies against Germany when it mobilized against Austria–Hungary, and that Germany also attack France and Belgium if it fought Russia. Barbara Tuchman writes that Europe was swept into war by "the pull of military schedules," and recalls Moltke's famous answer when the Kaiser asked if the German armies could be mobilized to the East: "Your Majesty, it cannot be done. The deployment of millions cannot be improvised. If Your Majesty insists on leading the whole army to the East it will not be an army ready for battle but a disorganized mob of armed men with no arrangements for supply."[120] Likewise, Herman Kahn notes the "rigid war plan[s]" of 1914, which "were literally cast in concrete,"[121] and David Ziegler notes the influence of military "planning in advance," which left "no time to improvise."[122]

The scope and character of these plans in turn reflected the assumption that the offense was strong. In an offense-dominant world Russia would have been prudent to mobilize against Germany if it mobilized against Austria–Hungary; and Germany probably would have been prudent to attack Belgium and France at the start of any Russo–German war. Thus the troublesome railroad schedules of 1914 reflected the offense-dominant world in which the schedulers believed they lived. Had they known that the defense was powerful, they would have been drawn towards flexible plans for limited deployment on single frontiers; and had such planning prevailed, the war might have been confined to Eastern Europe or the Balkans.

Moreover, the "inflexibility" of the war plans may have reflected the same offensive assumptions which determined their shape. Russian and German soldiers understandably developed only options which they believed prudent to exercise, while omitting plans which they believed would be dangerous to implement. These judgments in turn reflected their own and their adver-

---

120. Tuchman, *Guns of August*, pp. 92, 99.
121. Kahn, *On Thermonuclear War*, pp. 359, 362.
122. David W. Ziegler, *War, Peace and International Politics* (Boston: Little, Brown, 1977), p. 25.

saries' offensive ideas. Options were few because these offensive ideas seemed to narrow the range of prudent choice.

Lastly, the assumption of offense-dominance gave preset plans greater influence over the conduct of the July crisis, by raising the cost of improvisation if statesmen insisted on adjusting plans at the last minute. Russian statesmen were told that an improvised partial mobilization would place Russia in a "extremely dangerous situation,"[123] and German civilians were warned against improvisation in similar terms. This in turn reflected the size of the "windows" which improvised partial mobilizations would open for the adversary on the frontier which the partial mobilization left unguarded, which in turn reflected the assumption that the offense was strong (since if defenses were strong a bungled mobilization would create less opportunity for others to exploit). Thus the cult of the offensive gave planners greater power to bind statesmen to the plans they had prepared.

RUSSIAN MOBILIZATION PLANS. On July 28, 1914, Russian leaders announced that partial Russian mobilization against Austria would be ordered on July 29. They took this step to address threats emanating from Austria, acting partly to lend emphasis to their warnings to Austria that Russia would fight if Serbia were invaded, partly to offset Austrian mobilization against Serbia, and partly to offset or forestall Austrian mobilization measures which they believed were taking place or which they feared might eventually take place against Russia in Galicia.[124] However, after this announcement was made, Russian military officers advised their civilian superiors that no plans for partial mobilization existed, that such a mobilization would be a "pure improvisation," as General Denikin later wrote, and that sowing confusion in the Russian railway timetables would impede Russia's ability to mobilize later on its northern frontier. General Sukhomlinov warned the Czar that "much time would be necessary in which to re-establish the normal conditions for any further mobilization" following a partial mobilization, and General Yanushkevich flatly told Sazonov that general mobilization "could not be put into operation" once partial mobilization began.[125] Thus Russian lead-

123. By Generals Yanushkevich and Sukhomlinov, according to Sazonov, quoted in Albertini, *Origins*, Vol. 2, p. 566. See also M.F. Schilling, "Introduction," in *How the War Began*, trans. W. Cyprian Bridge, with a Foreword by S.D. Sazonov (London: Allen & Unwin, 1925), pp. 16, 63.
124. On the Russian decision, see Schmitt, *Coming of the War*, Vol. 2, pp. 85–87, 94–101; and Albertini, *Origins*, Vol. 2, pp. 539–561.
125. Anton I. Denikin, *The Career of a Tsarist Officer: Memoirs, 1872–1916*, trans. Margaret Patoski (Minneapolis: University of Minnesota Press, 1975), p. 222; Albertini, *Origins*, Vol. 2, p. 559; Schilling, *How the War Began*, p. 16.

ers were forced to choose between full mobilization or complete retreat, choosing full mobilization on July 30.

The cult of the offensive set the stage for this decision by buttressing Russian military calculations that full mobilization was safer than partial. We have little direct evidence explaining why Russian officers had prepared no plan for partial mobilization, but we can deduce their reasoning from their opinions on related subjects. These suggest that Russian officers believed that Germany would attack Russia if Russia fought Austria, and that the side mobilizing first would have the upper hand in a Russo–German war (as I have outlined above). Accordingly, it followed logically that Russia should launch any war with Austria by preempting Germany.

Russian leaders had three principal reasons to fear that Germany would not stand aside in an Austro–Russian conflict. First, the Russians were aware of the international Social Darwinism then sweeping Germany, and the expansionist attitude toward Russia which this worldview engendered. One Russian diplomat wrote that Germany was "beating all records of militarism" and "The Germans are not . . . wholly without the thought of removing from Russia at least part of the Baltic coastline in order to place us in the position of a second Serbia" in the course of a campaign for "German hegemony on the continent."[126] Russian military officers monitored the bellicose talk across the border with alarm, one intelligence report warning: "In Germany at present, the task of gradually accustoming the army and the population to the thought of the inevitability of conflict with Russia has begun," noting the regular public lectures which were then being delivered in Germany to foster war sentiment.[127]

Second, the Russians were aware of German alarm about windows and the talk of preventive war which this alarm engendered in Germany. Accordingly, Russian leaders expected that Germany might seize the excuse offered by a Balkan war to mount a preventive strike against Russia, especially since a war arising from the Balkans was a "best case" scenario for Germany, involving Austria on the side of Germany as it did. Thus General Yanushkevich explained Russia's decision to mobilize against Germany in 1914: "We knew well that Germany was ready for war, that she was longing

126. G.N. Trubetskoy, in 1909, quoted in Lieven, *Russia & the Origins of the First World War*, p. 96.

127. The Kiev District Staff, February 23, 1914, quoted in Fuller, "The Russian Empire and Its Potential Enemies," p. 17.

for it at that moment, because our big armaments program was not yet completed . . . and because our war potential was not as great as it might be." Accordingly, Russia had to expect war with Germany: "We knew that war was inevitable, not only against Austria, but also against Germany. For this reason partial mobilization against Austria alone, which would have left our front towards Germany open . . . might have brought about a disaster, a terrible disaster."[128] In short, Russia had to strike to preempt a German preventive strike against Russia.

Third, the Russians knew that the Germans believed that German and Austrian security were closely linked. Germany would therefore feel compelled to intervene in any Austro–Russian war, because a Russian victory against Austria would threaten German safety. German leaders had widely advertised this intention: for instance, Bethmann Hollweg had warned the Reichstag in 1912 that if the Austrians "while asserting their interests should against all expectations be attacked by a third party, then we would have to come resolutely to their aid. And then we would fight for the maintenance of our own position in Europe and in defense of our future and security."[129] And in fact this was precisely what happened in 1914: Germany apparently decided to attack on learning of Russian *partial* mobilization, before Russian full mobilization was known in Germany.[130] This suggests that the role of "inflexible" Russian plans in causing the war is overblown—Russian full mobilization was sufficient but not necessary to cause the war; but it also helps explain why these plans were drawn as they were, and supports the view that some of the logic behind them was correct, given the German state of mind with which Russia had to contend.

---

128. Albertini, *Origins*, Vol. 2, p. 559. See also Fuller, "The Russian Empire and Its Potential Enemies," p. 16.
129. In 1912, quoted in Stern, *Failure of Illiberalism*, p. 84. Likewise the Kaiser explained that security requirements compelled Germany to defend Austria: "If we are forced to take up arms it will be to help *Austria*, not only to defend ourselves against Russia but against the Slavs in general and to remain *German*. . . ." In 1912, quoted in Fischer, *War of Illusions*, pp. 190–191, emphasis in original. The German White Book also reflected this thinking, declaring that the "subjugation of all the Slavs under Russian sceptre" would render the "position of the Teutonic race in Central Europe untenable." August 3, 1914, quoted in Geiss, *German Foreign Policy*, p. 172.
130. See Schmitt, *Coming of the War*, Vol. 2, pp. 198–199; and Albertini, *Origins*, Vol. 3, pp. 7, 17–27; also Vol. 2, p. 485n. As Jagow plainly told the Russians on July 29: "If once you mobilize against Austria, then you will also take serious measures against us. . . . We are compelled to proclaim mobilization against Russia. . . ." Schmitt, *Coming of the War*, Vol. 2, p. 140.

In sum, Russians had to fear that expansionist, preventive, and alliance concerns might induce Germany to attack, which in turn reflected the German assumption that the offense was strong. The Russian belief that it paid to mobilize first reflected the effects of the same assumption in Russia. Had Europe known that the defense dominated, Russians would have had less reason to fear that an Austro–Russian war would spark a German attack, since the logic of expansionism and preventive war would presumably have been weaker in Germany, and Germany could more easily have tolerated some reduction in Austrian power without feeling that German safety was also threatened. At the same time, Russian soldiers would presumably have been slower to assume that they could improve their position in a Russo–German war by mobilizing preemptively. In short, the logic of general mobilization in Russia largely reflected and depended upon conclusions deduced from the cult of the offensive, or from its various manifestations. Without the cult of the offensive, a partial southern mobilization would have been the better option for Russia.

It also seems probable that the same logic helped persuade the Russian General Staff to eschew planning for a partial mobilization. If circumstances argued against a partial mobilization, they also argued against planning for one, since this would raise the risk that Russian civilians might actually implement the plan. This interpretation fits with suggestions that Russian officers exaggerated the difficulties of partial mobilization in their representations to Russian civilians.[131] If Russian soldiers left a partial mobilization option undeveloped because they believed that it would be dangerous to exercise, it follows that they also would emphasize the difficulty of improvising a southern option, since they also opposed it on other grounds.

GERMAN MOBILIZATION PLANS. The Schlieffen Plan was a disastrous scheme which only approached success because the French war plan was equally foolish: had the French army stood on the defensive instead of lunging into Alsace–Lorraine, it would have smashed the German army at the French frontier. Yet General Schlieffen's plan was a sensible response to the offense-

---

131. See L.C.F. Turner, "The Russian Mobilization in 1914," *Journal of Contemporary History*, Vol. 3, No. 1 (January 1968), pp. 72–74. But see also Lieven, *Russia and the Origins of the First World War*, pp. 148–150.

Likewise, German soldiers exaggerated the difficulties of adapting to eastward mobilization, as many observers note, e.g., Tuchman, *Guns of August*, p. 100, and Lebow, *Between Peace and War*, p. 236.

dominant world imagined by many Germans. The plan was flawed because it grew from a fundamentally flawed image of warfare.

In retrospect, Germany should have retained the later war plan of the elder Moltke (Chief of Staff from 1857 to 1888), who would have conducted a limited offensive in the east against Russia while standing on the defensive in the west.[132] However, several considerations pushed German planners instead toward Schlieffen's grandiose scheme, which envisioned a quick victory against Belgium and France, followed by an offensive against Russia.

First, German planners assumed that France would attack Germany if Germany fought Russia, leaving Germany no option for a one-front war. By tying down German troops in Poland, an eastern war would create a yawning window of opportunity for France to recover its lost territories, and a decisive German victory over Russia would threaten French security by leaving France to face Germany alone. For these reasons they believed that France would be both too tempted and too threatened to stand aside. Bernhardi, among others, pointed out "the standing danger that France will attack us on a favorable occasion, as soon as we find ourselves involved in complications elsewhere."[133] The German declaration of war against France explained that France might suddenly attack from behind if Germany fought Russia; hence, "Germany cannot leave to France the choice of the moment" at which to attack.[134]

Second, German planners assumed that "window" considerations required a German offensive against either France or Russia at the outset of any war against the Entente. German armies could mobilize faster than the combined Entente armies; hence, the ratio of forces would most favor Germany at the beginning of the war. Therefore, Germany would do best to force an early decision, which in turn required that it assume the offensive, since otherwise its enemies would play a waiting game. As one observer explained, Germany

---

132. Assessing the Schlieffen Plan are Ritter, *The Schlieffen Plan,* and Snyder, "Defending the Offensive," pp. 189–294.
133. Bernhardi, quoted in Anon., *Germany's War Mania* (London: A.W. Shaw, 1914), p. 161.
134. Albertini, *Origins,* Vol. 3, p. 194. Moreover, these fears reflected views found in France. When Poincaré was asked on July 29 if he believed war could be avoided, he reportedly replied: "It would be a great pity. We should never again find conditions better." Albertini, *Origins,* Vol. 3, p. 82n. Likewise, in 1912 the French General Staff concluded that a general war arising from the Balkans would leave Germany "at the mercy of the Entente" because Austrian forces would be diverted against Serbia, and "the Triple Entente would have the best chances of success and might gain a victory which would enable the map of Europe to be redrawn." Turner, *Origins,* p. 36. See also the opinions of Izvolsky and Bertie in Schmitt, *Coming of the War,* Vol. 1, pp. 20–21, and Vol. 2, p. 349n.

"has the speed and Russia has the numbers, and the safety of the German Empire forbade that Germany should allow Russia time to bring up masses of troops from all parts of her wide dominions."[135] Germans believed that the window created by these differential mobilization rates was big, in turn, because they believed that both Germany and its enemies could mount a decisive offensive against the other with a small margin of superiority. If Germany struck at the right time, it could win easily—Germans hoped for victory in several weeks, as noted above—while if it waited it was doomed by Entente numerical superiority, which German defenses would be too weak to resist.

Third, German planners believed that an offensive against France would net them more than an offensive against Russia, which explains the western bias of the Schlieffen Plan. France could be attacked more easily than Russia, because French forces and resources lay within closer reach of German power; hence, as Moltke wrote before the war, "A speedy decision may be hoped for [against France], while an offensive against Russia would be an interminable affair."[136] Moreover, France was the more dangerous opponent not to attack, because it could take the offensive against Germany more quickly than Russia, and could threaten more important German territories if Germany left its frontier unguarded. Thus Moltke explained that they struck westward because "Germany could not afford to expose herself to the danger of attack by strong French forces in the direction of the Lower Rhine," and Wegerer wrote later that the German strike was compelled by the need to protect the German industrial region from French attack.[137] In German eyes these considerations made it too dangerous to stand on the defensive in the West in hopes that war with France could be avoided.

Finally, German planners believed that Britain would not have time to bring decisive power to bear on the continent before the German army overran France. Accordingly, they discounted the British opposition which their attack on France and Belgium would elicit: Schlieffen declared that if the British army landed, it would be "securely billeted" at Antwerp or "arrested" by the German armies,[138] while Moltke said he hoped that it would

135. Goschen, in Schmitt, *Coming of the War*, Vol. 2, p. 321.
136. Moltke, in General Ludendorff, *The General Staff and its Problems*, trans. F.A. Holt (New York: E.P. Dutton, n.d.), Vol. 1, p. 61.
137. Geiss, *July 1914*, p. 357; and Alfred von Wegerer, *A Refutation of the Versailles War Guilt Thesis*, trans. Edwin H. Zeydel (New York: Alfred A. Knopf, 1930), p. 310.
138. Ritter, *Schlieffen Plan*, pp. 71, 161–162; and Geiss, *German Foreign Policy*, p. 101. See also Ritter, *Schlieffen Plan*, p. 161. But see also Moltke quoted in Turner, *Origins of the World War*, p. 64.

land so that the German army "could take care of it."[139] In accordance with their "bandwagon" worldview, German leaders also hoped that German power might cow Britain into neutrality; or that Britain might hesitate before entering the war, and then might quit in discouragement once the French were beaten—Schlieffen expected that, "If the battle [in France] goes in favor of the Germans, the English are likely to abandon their enterprise as hopeless"—which led them to further discount the extra political costs of attacking westward.[140]

Given these four assumptions, an attack westward, even one through Belgium which provoked British intervention, was the most sensible thing for Germany to do. Each assumption, in turn, was a manifestation of the belief that the offense was strong. Thus while the Schlieffen Plan has been widely criticized for its political and military naiveté, it would have been a prudent plan had Germans actually lived in the offense-dominant world they imagined. Under these circumstances quick mobilization would have in fact given them a chance to win a decisive victory during their window of opportunity, and if they had failed to exploit this window by attacking, they would eventually have lost; the risk of standing on the defense in the West in hopes that France would not fight would have been too great; and the invasion of France and Belgium would have been worth the price, because British power probably could not have affected the outcome of the war.

Thus the belief in the power of the offense was the linchpin which held Schlieffen's logic together, and the main criticisms which can be levelled at the German war plan flow from the falsehood of this belief. German interests would have been better served by a limited, flexible, east-only plan which conformed to the defensive realities of 1914. Moreover, had Germany adopted such a plan, the First World War might well have been confined to Eastern Europe, never becoming a world war.

"MOBILIZATION MEANS WAR"

"Mobilization meant war" in 1914 because mobilization meant war to Germany: the German war plan mandated that special units of the German standing army would attack Belgium and Luxemburg immediately after mobilization was ordered, and long before it was completed. (In fact Germany

---

139. Ritter, *Sword and the Scepter*, Vol. 2, p. 157.
140. Ritter, *The Schlieffen Plan*, p. 163. See also Bethmann Hollweg, quoted in Fischer, *War of Illusions*, pp. 169, 186–187.

invaded Luxemburg on August 1, the same day on which it ordered full mobilization.) Thus Germany had no pure "mobilization" plan, but rather had a "mobilization and attack" plan under which mobilizing and attacking would be undertaken simultaneously. As a result, Europe would cascade into war if any European state mobilized in a manner which eventually forced German mobilization.

This melding of mobilization and attack in Germany reflected two decisions to which I have already alluded. First, Germans believed that they would lose their chance for victory and create a grave danger for themselves if they gave the Entente time to mobilize its superior numbers. In German eyes, German defenses would be too weak to defeat this superiority. As one German apologist later argued, "Germany could never with success have warded off numerically far superior opponents by means of a defensive war against a mobilized Europe" had it mobilized and stood in place. Hence it was "essential for the Central Powers to begin hostilities as soon as possible" following mobilization.[141] Likewise, during the July crisis, Jagow explained that Germany must attack in response to Russian mobilization because "we are obliged to act as fast as possible before Russia has the time to mobilize her army."[142]

Second, the German war plan depended on the quick seizure of Liège. Germany could only secure Liège quickly if German troops arrived before Belgium prepared its defense, and this in turn depended on achieving surprise against Belgium. Accordingly, German military planners enshrouded the planned Liège attack in such dark secrecy that Bethmann Hollweg, Admiral Tirpitz, and possibly even the Kaiser were unaware of it.[143] They also felt compelled to strike as soon as mobilization was authorized, both because Belgium would strengthen the defenses of Liège as a normal part of the Belgian mobilization which German mobilization would engender, and because otherwise Belgium eventually might divine German intentions towards Liège and focus upon preparing its defense and destroying the critical bridges and tunnels which it controlled.

---

141. Von Wegerer, *Refutation*, pp. 307–309.
142. August 4, quoted in Alfred Vagts, *Defense and Diplomacy* (New York: Kings Crown Press, 1956), p. 306. Likewise Bethmann Hollweg explained that, if Russia mobilized, "we could hardly sit and talk any longer because we have to strike immediately in order to have any chance of winning at all." Fischer, *War of Illusions*, p. 484.
143. Albertini, *Origins*, Vol. 2, p. 581; Vol. 3, pp. 195, 250, 391; Ritter, *Sword and the Scepter*, Vol. 2, p. 266; and Fay, *Origins*, Vol. 1, pp. 41–42.

Both of these decisions in turn reflected German faith in the power of the offense, and were not appropriate to a defense-dominant world. Had Germans recognized the actual power of the defense, they might have recognized that neither Germany nor its enemies could win decisively even by exploiting a fleeting material advantage, and decided instead to mobilize without attacking. The tactical windows that drove Germany to strike in 1914 were a mirage, as events demonstrated during 1914–1918, and Germans would have known this in advance had they understood the power of the defense. Likewise, the Liège *coup de main* was an artifact of Schlieffen's offensive plan; if the Germans had stuck with the elder Moltke's plan, they could have abandoned both the Liège attack and the compulsion to strike quickly which it helped to engender.

BRINKMANSHIP AND FAITS ACCOMPLIS

Two *faits accomplis* by the Central powers set the stage for the outbreak of the war: the Austrian ultimatum to Serbia on July 23, and the Austrian declaration of war against Serbia on July 28. The Central powers also planned to follow these with a third *fait accompli,* by quickly smashing Serbia on the battlefield before the Entente could intervene. These plans and actions reflected the German strategy for the crisis: "*fait accompli* and then friendly towards the Entente, the shock can be endured," as Kurt Riezler had summarized.[144]

This *fait accompli* strategy deprived German leaders of warning that their actions would plunge Germany into a world war, by depriving the Entente of the chance to warn Germany that it would respond if Austria attacked Serbia. It also deprived diplomats of the chance to resolve the Austro–Serbian dispute in a manner acceptable to Russia. Whether this affected the outcome of the crisis depends on German intentions—if Germany sought a pretext for a world war, then this missed opportunity had no importance, but if it preferred the status quo to world war, as I believe it narrowly did, then the decision to adopt *fait accompli* tactics was a crucial step on the road to war.

---

144. July 8, quoted in John A. Moses, *The Politics of Illusion: The Fischer Controversy in German Historiography* (London: George Prior, 1975), p. 39. Austria declared war on Serbia, as one German diplomat explained, "in order to forestall any attempt at mediation" by the Entente; and the rapid occupation of Serbia was intended to "confront the world with a '*fait accompli.*'" Tschirschky, in Schmitt, *Coming of the War,* Vol. 2, p. 5; and Jagow, in Albertini, *Origins,* Vol. 2, p. 344; see also pp. 453–460.

Had Germany not done so, it might have recognized where its policies led before it took irrevocable steps, and have drawn back.

The influence of the cult of the offensive is seen both in the German adoption of this *fait accompli* strategy and in the disastrous scope of the results which followed in its train. Some Germans, such as Kurt Riezler, apparently favored brinkmanship and *fait accompli* diplomacy as a means of peaceful expansion.[145] Others probably saw it as a means to provoke a continental war. In either case it reflected a German willingness to trade peace for territory, which reflected German expansionism—which in turn reflected security concerns fuelled by the cult of the offensive. Even those who saw *faits accomplis* as tools of peaceful imperialism recognized their risks, believing that necessity justified the risk. Thus Riezler saw the world in Darwinistic terms: "each people wants to grow, expand, dominate and subjugate others without end . . . until the world has become an organic unity under [single] domination."[146] *Faits accomplis* were dangerous tools whose adoption reflected the dangerous circumstances which Germans believed they faced.

The cult of the offensive also stiffened the resistance of the Entente to the Austro–German *fait accompli,* by magnifying the dangers they believed it posed to their own security.[147] Thus Russian leaders believed that Russian security would be directly jeopardized if Austria crushed Serbia, because they valued the power which Serbia added to their alliance, and because they feared a domino effect, running to Constantinople and beyond, if Serbia were overrun. Sazonov believed that Serbian and Bulgarian military power was a vital Russian resource, "five hundred thousand bayonets to guard the Balkans" which "would bar the road forever to German penetration, Austrian invasion."[148] If this asset were lost, Russia's defense of its own territories would be jeopardized by the German approach to Constantinople: Sazonov warned the Czar, "First Serbia would be gobbled up; then will come Bulgaria's turn, and then we shall have her on the Black Sea." This would be

---

145. On Riezler's thought, see Moses, *Politics of Illusion,* pp. 27–44; and Thompson, *In the Eye of the Storm.*
146. Quoted in Moses, *Politics of Illusion,* pp. 28, 31. Likewise during the war Riezler wrote that unless Germany gained a wider sphere of influence in Europe "we will in the long run be crushed between the great world empires . . . Russia and England." Thompson, *In the Eye of the Storm,* p. 107.
147. I am grateful to Jack Snyder for this and related observations.
148. Schmitt, *Coming of the War,* Vol. 1, p. 131n. See also Lieven, *Russia and the Origins of the First World War,* pp. 40–41, 99–100, 147.

"the death-warrant of Russia" since in such an event "the whole of southern Russia would be subject to [Germany]."[149]

Similar views could be found in France. During the July crisis one French observer warned that French and Serbian security were closely intertwined, and the demise of Serbia would directly threaten French security:

> To do away with Serbia means to double the strength which Austria can send against Russia: to double Austro–Hungarian resistance to the Russian Army means to enable Germany to send some more army corps against France. For every Serbian soldier killed by a bullet on the Morava one more Prussian soldier can be sent to the Moselle. . . . It is for us to grasp this truth and draw the consequences from it before disaster overtakes Serbia.[150]

These considerations helped spur the Russian and French decisions to begin military preparations on July 25, which set in train a further sequence of events: German preliminary preparations, which were detected and exaggerated by French and Russian officials, spurring them on to further measures, which helped spur the Germans to their decision to mobilize on July 30. The effects of the original *fait accompli* rippled outward in ever-wider circles, because the reactions of each state perturbed the safety of others—forcing them to react or preempt, and ultimately forcing Germany to launch a world war which even it preferred to avoid.

Had Europe known that, in reality, the defense dominated, these dynamics might have been dampened: the compulsion to resort to *faits accomplis*, the scope of the dangers they raised for others, and the rippling effects engendered by others' reactions all would have been lessened. States still might have acted as they did, but they would have been less pressured in this direction.

PROBLEMS OF ALLIANCES: UNCONDITIONALITY AND AMBIGUITY

Two aspects of the European alliance system fostered the outbreak of World War I and helped spread the war. First, both alliances had an unconditional, offensive character—allies supported one another unreservedly, regardless of whether their behavior was defensive or provocative. As a result a local war would tend to spread throughout Europe. And second, German leaders were not convinced that Britain would fight as an Entente member, which

---

149. Fay, *Origins*, Vol. 2, p. 300; Sazonov, *Fateful Years*, p. 179; Schmitt, *Coming of the War*, Vol. 1, p. 87.
150. J. Herbette, July 29, in Albertini, *Origins*, Vol. 2, p. 596.

encouraged Germany to confront the Entente. In both cases the cult of the offensive contributed to the problem.

UNCONDITIONAL ("TIGHT") ALLIANCES. Many scholars contend that the mere existence of the Triple Alliance and the Triple Entente caused and spread the war. Sidney Fay concluded, "The greatest single underlying cause of the War was the system of secret alliance," and Raymond Aron argued that the division of Europe into two camps "made it inevitable that any conflict involving two great powers would bring general war."[151] But the problem with the alliances of 1914 lay less with their existence than with their nature. A network of defensive alliances, such as Bismarck's alliances of the 1880s, would have lowered the risk of war by facing aggressors with many enemies, and by making status quo powers secure in the knowledge that they had many allies. Wars also would have tended to remain localized, because the allies of an aggressor would have stood aside from any war that aggressor had provoked. Thus the unconditional nature of alliances rather than their mere existence was the true source of their danger in 1914.

The Austro–German alliance was offensive chiefly and simply because its members had compatible aggressive aims. Moreover, German and Russian mobilization plans left their neighbors no choice but to behave as allies by putting them all under threat of attack. But the Entente also operated more unconditionally, or "tightly," because Britain and France failed to restrain Russia from undertaking mobilization measures during the July crisis. This was a failure in alliance diplomacy, which in turn reflected constraints imposed upon the Western allies by the offensive assumptions and preparations with which they had to work.

First, they were hamstrung by the offensive nature of Russian military doctrine, which left them unable to demand that Russia confine itself to defensive preparations. All Russian preparations were inherently offensive, because Russian war plans were offensive. This put Russia's allies in an "all or nothing" situation—either they could demand that Russia stand unprepared, or they could consent to provocative preparations. Thus the British ambassador to St. Petersburg warned that Britain faced a painful decision, to "choose between giving Russia our active support or renouncing her friendship."[152] Had Russia confined itself to preparing its own defense, it

151. Fay, *Origins*, Vol. 1, p. 34; and Raymond Aron, *The Century of Total War* (Boston: Beacon Press, 1955), p. 15.
152. Buchanan, in Fay, *Origins*, Vol. 2, p. 379.

would have sacrificed its Balkan interests by leaving Austria free to attack Serbia, and this it would have been very reluctant to do. However, the British government was probably willing to sacrifice Russia's Balkan interests to preserve peace;[153] what Britain was unable to do was to frame a request to Russia which would achieve this, because there was no obvious class of defensive activity that it could demand. Edward Grey, the British Foreign Secretary, wrote later:

> I felt impatient at the suggestion that it was for me to influence or restrain Russia. I could do nothing but express pious hopes in general terms to Sazonov. If I were to address a direct request to him that Russia should not mobilize, I knew his reply: Germany was much more ready for war than Russia; it was a tremendous risk for Russia to delay her mobilization. . . . I did most honestly feel that neither Russian nor French mobilization was an unreasonable or unnecessary precaution.[154]

One sees in this statement a losing struggle to cope with the absence of defensive options. Russia was threatened, and must mobilize. How could Britain object?

Britain and France were also constrained by their dependence upon the strength and unity of the Entente for their own security, which limited their ability to make demands on Russia. Because they feared they might fracture the Entente if they pressed Russia too hard, they tempered their demands to preserve the alliance. Thus Poincaré wrote later that France had been forced to reconcile its efforts to restrain Russia with the need to preserve the Franco–Russian alliance, "the break up of which would leave us in isolation at the mercy of our rivals."[155] Likewise Winston Churchill recalled that "the one thing [the Entente states] would not do was repudiate each other. To do this might avert the war for the time being. It would leave each of them to face the next crisis alone. They did not dare to separate."[156] These fears were probably overdrawn, since Russia had no other option than alliance with the other Entente states, but apparently they affected French and British behavior.[157] This in turn reflected the assumption in France and Britain that the security of the Entente members was closely interdependent.

---

153. See Geiss, *July 1914*, p. 176; and Albertini, *Origins*, Vol. 2, p. 295.
154. Albertini, *Origins*, Vol. 2, p. 518.
155. Ibid., p. 605.
156. Winston Churchill, *The Unknown War* (New York: Charles Scribner's Sons, 1931), p. 103.
157. Thus Grey later wrote that he had feared a "diplomatic triumph on the German side and humiliation on the other as would smash the Entente, and if it did not break the Franco–Russian

French leaders also felt forced in their own interests to aid Russia if Russia embroiled itself with Germany, because French security depended on the maintenance of Russian power. This in turn undermined the French ability to credibly threaten to discipline a provocative Russia. Thus the British ambassador to Paris reflected French views when he cabled that he could not imagine that France would remain quiescent during a Russo–German war, because "If [the] French undertook to remain so, the Germans would first attack [the] Russians and, if they defeated them, they would then turn round on the French."[158] This prospect delimited French power to restrain Russian conduct.

Third, British leaders were unaware that German mobilization meant war, hence that peace required Britain to restrain Russia from mobilizing first, as well as attacking. As a result, they took a more relaxed view of Russian mobilization than they otherwise might, while frittering away their energies on schemes to preserve peace which assumed that war could be averted even after the mobilizations began.[159] This British ignorance reflected German failure to explain clearly to the Entente that mobilization did indeed mean war—German leaders had many opportunities during the July crisis to make this plain, but did not do so.[160] We can only guess why Germany was silent, but German desire to avoid throwing a spotlight on the Liège operation probably played a part, leading German soldiers to conceal the plan from German civilians, which led German civilians to conceal the political implications of the plan from the rest of Europe.[161] Thus preemptive planning threw a shroud of secrecy over military matters, which obscured the mechanism that would unleash the war and rendered British statesmen less able

---

alliance, would leave it without spirit, a spineless and helpless thing." Likewise during July 1914 Harold Nicolson wrote: "Our attitude during the crisis will be regarded by Russia as a test and we must be careful not to alienate her." Schmitt, *Coming of the War,* Vol. 2, pp. 38, 258.

158. Bertie, on August 1, in Schmitt, *Coming of the War,* Vol. 2, p. 349n.

159. Geiss, *July 1914,* pp. 198, 212–213, 250–251; Albertini, *Origins,* Vol. 2, pp. 330–336.

160. See Albertini, *Origins,* Vol. 2, pp. 479–481; Vol. 3, pp. 41–43, 61–65. Albertini writes that European leaders "had no knowledge of what mobilization actually was . . . what consequences it brought with it, to what risks it exposed the peace of Europe. They looked on it as a measure costly, it is true, but to which recourse might be had without necessarily implying that war would follow." This reflected German policy: Bethmann's ultimatum to Russia "entirely omitted to explain that for Germany to mobilize meant to begin war," and Sazonov gathered "the distinct impression that German mobilization was not equivalent to war" from his exchanges with German officials. Vol. 2, p. 479; Vol. 3, pp. 41–43.

161. Kautsky and Albertini suggest that the German deception was intended to lull the Russians into military inaction, but it seems more likely that they sought to lull the Belgians. Albertini, *Origins,* Vol. 3, p. 43.

to wield British power effectively for peace by obscuring what it was that Britain had to do.

Lastly, the nature of German war plans empowered Russia to involve France, and probably Britain also, in war, since Germany would be likely to start any eastern war by attacking westward, as Russian planners were aware. Hence France and Britain would probably have to fight for Russia even if they preferred to stand aside, because German planners assumed that France would fight eventually and planned accordingly, and the plans they drew would threaten vital British interests. We have no direct evidence that Russian policies were emboldened by these considerations, but it would be surprising if they never occurred to Russian leaders.

These dynamics reflected the general tendency of alliances toward tightness and offensiveness in an offense-dominant world. Had Europe known that the defense had the advantage, the British and French could have more easily afforded to discipline Russia in the interest of peace, and this might have affected Russian calculations. Had Russia had a defensive military strategy, its allies could more easily and legitimately have asked it to confine itself to defensive preparations. Had British leaders better understood German war plans, they might have known to focus their efforts on preventing Russian mobilization. And had German plans been different, Russian leaders would have been more uncertain that Germany would entangle the Western powers in eastern wars, and perhaps proceeded more cautiously.

The importance of the failure of the Western powers to restrain Russia can be exaggerated, since Russia was not the chief provocateur in the July crisis. Moreover, too much can be made of factors which hamstrung French restraint of Russia, since French desire to prevent war was tepid at best, so French inaction probably owed as much to indifference as inability. Nevertheless, Russian mobilization was an important step toward a war which Britain, if not France, urgently wanted to prevent; hence, to that extent, the alliance dynamics which allowed it helped bring on the war.

THE AMBIGUITY OF BRITISH POLICY. The British government is often accused of causing the war by failing to warn Germany that Britain would fight. Thus Albertini concludes that "to act as Grey did was to allow the catastrophe to happen,"[162] and Germans themselves later argued that the British had led them on, the Kaiser complaining of "the grossest deception" by the British.[163]

162. Ibid., Vol. 2, p. 644.
163. Ibid., p. 517. See also Tirpitz, quoted in ibid., Vol. 3, p. 189.

The British government indeed failed to convey a clear threat to the Germans until after the crisis was out of control, and the Germans apparently were misled by this. Jagow declared on July 26 that "we are sure of England's neutrality," while during the war the Kaiser wailed, "If only someone had told me beforehand that England would take up arms against us!"[164] However, this failure was not entirely the fault of British leaders; it also reflected their circumstances. First, they apparently felt hamstrung by the lack of a defensive policy option. Grey voiced fear that if he stood too firmly with France and Russia, they would grow too demanding, while Germany would feel threatened, and "Such a menace would but stiffen her attitude."[165]

Second, British leaders were unaware of the nature of the German policy to which they were forced to react until very late, which left them little time in which to choose and explain their response. Lulled by the Austro–German *fait accompli* strategy, they were unaware until July 23 that a crisis was upon them. On July 6, Arthur Nicolson, undersecretary of the British foreign office, cheerfully declared, "We have no very urgent and pressing question to preoccupy us in the rest of Europe."[166] They also were apparently unaware that a continental war would begin with a complete German conquest of Belgium, thanks to the dark secrecy surrounding the Liège operation. Britain doubtless would have joined the war even if Germany had not invaded Belgium, but the Belgian invasion provoked a powerful emotional response in Britain which spurred a quick decision on August 4. This reaction suggests that the British decision would have been clearer to the British, hence to the Germans, had the nature of the German operation been known in advance.

Thus the British failure to warn Germany was due as much to German secrecy as to British indecision. Albertini's condemnation of Grey seems unfair: governments cannot easily take national decisions for war in less than a week in response to an uncertain provocation. The ambiguity of British policy should be recognized as an artifact of the secret styles of the Central powers, which reflected the competitive politics and preemptive military doctrines of the times.

WHY SO MANY "BLUNDERS"?

Historians often ascribe the outbreak of the war to the blunders of a mediocre European leadership. Barbara Tuchman describes the Russian Czar as having

164. Ibid., Vol. 2, p. 429; and Tuchman, *Guns of August*, p. 143. See also Albertini, *Origins*, Vol. 2, pp. 514–527, 643–650; and Jarausch, "Illusion of Limited War."
165. Albertini, *Origins*, Vol. 2, p. 631; and Schmitt, *Coming of the War*, Vol. 2, p. 90.
166. Schmitt, *Coming of the War*, Vol. 1, pp. 417–418.

"a mind so shallow as to be all surface," and Albertini refers to the "untrained, incapable, dull-witted Bethmann-Hollweg," the "mediocrity of all the personages" in the German government, and the "short-sighted and unenlightened" Austrians. Ludwig Reiners devotes a chapter to "Berchtold's Blunders"; Michael Howard notes the "bland ignorance among national leaders" of defense matters; and Oron Hale claims that "the men who directed international affairs in 1914 were at the lowest level of competence and ability in several decades."[167]

Statesmen often did act on false premises or fail to anticipate the consequences of their actions during the July crisis. For instance, Russian leaders were initially unaware that a partial mobilization would impede a later general mobilization;[168] they probably exaggerated the military importance of mobilizing against Austria quickly;[169] they falsely believed Germany would acquiesce to their partial mobilization; they probably exaggerated the significance of the Austrian bombardment of Belgrade;[170] they falsely believed a general Russian mobilization could be concealed from Germany; and they mobilized without fully realizing that for Germany "mobilization meant war."[171]

German leaders encouraged Russia to believe that Germany would tolerate a partial Russian mobilization, and failed to explain to Entente statesmen that mobilization meant war, leading British and Russian leaders to assume that it did not.[172] They also badly misread European political sentiment, hoping that Italy, Sweden, Rumania, and even Japan would fight with the Central powers, and that Britain and Belgium would stand aside.[173] For their part, Britain and Italy failed to warn Germany of their policies; and Britain acquiesced to Russian mobilization, apparently without realizing that Russian mobilization meant German mobilization, which meant war. Finally, intelli-

---

167. Tuchman, *Guns of August,* p. 78; Albertini, *Origins,* Vol. 2, pp. 389, 436; Vol. 3, p. 253; Ludwig Reiners, *The Lamps Went Out in Europe* (New York: Pantheon, 1955), pp. 112–122; Howard quoted in Schelling, *Arms and Influence,* p. 243; and Oron J. Hale, *The Great Illusion: 1900–1914* (New York: Harper & Row, 1971), p. 285.
168. Albertini, *Origins,* Vol. 2, pp. 295–296.
169. See Turner, *Origins of the First World War,* pp. 92–93; Albertini, *Origins,* Vol. 2, p. 409; Vol. 3, pp. 230–231; but see also Lieven, *Russia and the Origins of the First World War,* pp. 148–149.
170. Reiners, *Lamps Went Out in Europe,* p. 135; and Albertini, *Origins,* Vol. 2, p. 553.
171. Albertini, *Origins,* Vol. 2, p. 574, 579–581; Vol. 3, pp. 56, 60–65.
172. Ibid., Vol. 2, pp. 332, 479–482, 485, 499–500, 550; Vol. 3, pp. 41–43, 61–65; Geiss, *July 1914,* pp. 245, 253, 266.
173. See Albertini, *Origins,* Vol. 2, pp. 334, 673, 678; Vol. 3, p. 233; Geiss, *July 1914,* pp. 226, 255, 302, 350–353; Schmitt, *Coming of the War,* Vol. 1, pp. 72–74, 322; Vol. 2, pp. 52–55, 149, 390n. Also relevant is Albertini, *Origins,* Vol. 2, pp. 308–309, 480, 541.

gence mistakes on both sides made matters worse. Russian leaders exaggerated German and Austrian mobilization measures, some German reports exaggerated Russian mobilizations, and French officials exaggerated German measures, which helped spur both sides to take further measures.[174]

What explains this plethora of blunders and accidents? Perhaps Europe was unlucky in the leaders it drew, but conditions in 1914 also made mistakes easy to make and hard to undo. Because secrecy was tight and *faits accomplis* were the fashion, facts were hard to acquire. Because windows were large and preemption was tempting, mistakes provoked rapid, dramatic reactions that quickly made the mistake irreversible. Statesmen seem like blunderers in retrospect partly because the international situation in 1914 was especially demanding and unforgiving of error. Historians castigate Grey for failing to rapidly take drastic national decisions under confusing and unexpected circumstances in the absence of domestic political consensus, and criticize Sazonov for his shaky grasp of military details on July 28 which no Russian civilian had had in mind five days earlier. The standard implicit in these criticisms is too stiff—statecraft seldom achieves such speed and precision. The blame for 1914 lies less with the statesmen of the times than with the conditions of the times and the severe demands these placed on statesmen.

BLAMECASTING

The explosive conditions created by the cult of the offensive made it easier for Germany to spark war without being blamed, by enabling that country to provoke its enemies to take defensive or preemptive steps which confused the question of responsibility for the war. German advocates of preventive war believed that Germany had to avoid blame for its outbreak, to preserve British neutrality and German public support for the war. Moreover, they seemed confident that the onus for war *could* be substantially shifted onto their opponents. Thus Moltke counselled war but warned that "the attack must be started by the Slavs,"[175] Bethmann Hollweg decreed that "we must

---

174. See generally Lebow, *Between Peace and War*, pp. 238–242; and Albertini, *Origins*, Vol. 3, pp. 67–68. For details on Russia see Albertini, *Origins*, Vol. 2, pp. 499, 545–546, 549, 566–567, 570–571, 576; Schmitt, *Coming of the War*, Vol. 2, pp. 97–98, 238, 244n.; Schilling, *How the War Began*, pp. 61–62; and Sazonov, *Fateful Years*, pp. 193, 199–200, 202–203. For details on France, see Joffre, *Personal Memoirs*, Vol. 1, pp. 117–128; and Albertini, *Origins*, Vol. 2, p. 647; Vol. 3, p. 67. On Germany see Trumpener, "War Premeditated?," pp. 73–74; Albertini, *Origins*, Vol. 2, pp. 529, 560, 637; Vol. 3, pp. 2–3, 6–9; and Geiss, *July 1914*, pp. 291–294.
175. In 1913, in Albertini, *Origins*, Vol. 2, p. 486.

give the impression of being forced into war,"[176] and Admiral von Müller summarized German policy during the July crisis as being to "keep quiet, letting Russia put herself in the wrong, but then not shying away from war."[177] "It is very important that we should appear to have been provoked" in a war arising from the Balkans, wrote Jagow, for "then—but probably only then—Britain can remain neutral."[178] And as the war broke out, von Müller wrote, "The mood is brilliant. The government has succeeded very well in making us appear as the attacked."[179]

These and other statements suggest an official German hope that German responsibility could be concealed. Moreover, whatever the source of this confidence, it had a sound basis in prevailing military conditions, which blurred the distinction between offensive and defensive conduct, and forced such quick reactions to provocation that the question of "who started it?" could later be obscured. Indeed, the German "innocence campaign" during and after the war succeeded for many years partly because the war developed from a rapid and complex chemistry of provocation and response which could easily be misconstrued by a willful propagandist or a gullible historian.[180] Defenders seemed like aggressors to the untrained eye, because all defended quickly and aggressively. Jack Snyder rightly points out elsewhere in this issue that German war plans were poorly adapted for the strategy of brinkmanship and peaceful expansion which many Germans pursued until 1914, but prevailing European military arrangements and beliefs also facilitated the deceptions in which advocates of preventive war believed Germany had to engage.

---

176. On July 27, 1914, in Fischer, *War of Illusions*, p. 486.
177. On July 27, in J.C.G. Röhl, "Admiral von Müller and the Approach of War, 1911–1914," *Historical Journal*, Vol. 12, No. 4 (1969), p. 669. In the same spirit, Bernhardi (who hoped for Russian rather than British neutrality) wrote before the war that the task of German diplomacy was to spur a French attack, continuing: "[W]e must not hope to bring about this attack by waiting passively. Neither France nor Russia nor England need to attack in order to further their interests. . . . [Rather] we must initiate an active policy which, without attacking France, will so prejudice her interests or those of England that both these States would feel themselves compelled to attack us. Opportunities for such procedures are offered both in Africa and in Europe. . . ." Bernhardi, *Germany and the Next War*, p. 280.
178. In 1913, in Fischer, *War of Illusions*, p. 212.
179. Röhl, "Admiral von Müller," p. 670.
180. On this innocence campaign, see Imanuel Geiss, "The Outbreak of the First World War and German War Aims," in Walter Laqueur and George L. Mosse, eds., *1914: The Coming of the First World War* (New York: Harper and Row, 1966), pp. 71–78.

Conclusion

The cult of the offensive was a major underlying cause of the war of 1914, feeding or magnifying a wide range of secondary dangers which helped pull the world to war. The causes of the war are often catalogued as an unrelated grab-bag of misfortunes which unluckily arose at the same time; but many shared a common source in the cult of the offensive, and should be recognized as its symptoms and artifacts rather than as isolated phenomena.

The consequences of the cult of the offensive are illuminated by imagining the politics of 1914 had European leaders recognized the actual power of the defense. German expansionists then would have met stronger arguments that empire was needless and impossible, and Germany could have more easily let the Russian military buildup run its course, knowing that German defenses could still withstand Russian attack. All European states would have been less tempted to mobilize first, and each could have tolerated more preparations by adversaries before mobilizing themselves, so the spiral of mobilization and counter-mobilization would have operated more slowly, if at all. If armies mobilized, they might have rushed to defend their own trenches and fortifications, instead of crossing frontiers, divorcing mobilization from war. Mobilizations could more easily have been confined to single frontiers, localizing the crisis. Britain could more easily have warned the Germans and restrained the Russians, and all statesmen could more easily have recovered and reversed mistakes made in haste or on false information. Thus the logic that led Germany to provoke the 1914 crisis would have been undermined, and the chain reaction by which the war spread outward from the Balkans would have been very improbable. In all likelihood, the Austro-Serbian conflict would have been a minor and soon-forgotten disturbance on the periphery of European politics.

This conclusion does not depend upon how one resolves the "Fischer controversy" over German prewar aims; while the outcome of the Fischer debate affects the *way* in which the cult caused the war, it does not affect the importance which the cult should be assigned. If one accepts the Fischer–Geiss–Röhl view that German aims were very aggressive, then one emphasizes the role of the cult in feeding German expansionism, German window thinking, and the German ability to catalyze a war while concealing responsibility for it by provoking a preemption by Germany's adversaries. If one believes that Germany was less aggressive, then one focuses on the role of the incentive to preempt in spurring the Russian and French decisions to mobilize, the nature of Russian and German mobilization plans, the British failure to restrain Russia and warn Germany, the scope and irreversibility of the effects of the Austro–German *fait accompli,* and the various other blunders of statesmen.[181] The cult of the offensive would play a different role in the history as taught by these two schools, but a central role in both.

The 1914 case thus supports Robert Jervis and other theorists who propose that an offense-dominant world is more dangerous, and warns both superpowers against the offensive ideas which many military planners in both countries favor. Offensive doctrines have long been dogma in the Soviet military establishment, and they are gaining adherents in the United States as well. This is seen in the declining popularity of the nuclear strategy of "assured destruction" and the growing fashionability of "counterforce" nuclear strategies,[182] which are essentially offensive in nature.[183]

The 1914 case bears directly on the debate about these counterforce strategies, warning that the dangers of counterforce include but also extend far beyond the well-known problems of "crisis instability" and preemptive war. If the superpowers achieved disarming counterforce capabilities, or if they believed they had done so, the entire political universe would be disturbed. The logic of self-protection in a counterforce world would compel much of the same behavior and produce the same phenomena that drove the world to war in 1914—dark political and military secrecy, intense competition for resources and allies, yawning windows of opportunity and vulnerability, intense arms-racing, and offensive and preemptive war plans of great scope and violence. Smaller political and military mistakes would have larger and less reversible consequences. Crises would be harder to control, since military

---

181. A useful review of the debate about German aims is Moses, *Politics of Illusion.*
182. On the growth of offensive ideas under the Reagan Administration, see Barry R. Posen and Stephen Van Evera, "Defense Policy and the Reagan Administration: Departure from Containment," *International Security,* Vol. 8, No. 1 (Summer 1983), pp. 24–30. On counterforce strategies, a recent critical essay is Robert Jervis, *The Illogic of American Nuclear Strategy* (Ithaca: Cornell University Press, 1984).
183. "Counterforce" forces include forces which could preemptively destroy opposing nuclear forces before they are launched, forces which could destroy retaliating warheads in flight towards the attacker's cities, and forces which could limit the damage which retaliating warheads could inflict on the attacker's society if they arrived. Hence, "counterforce" weapons and programs include highly accurate ICBMs and SLBMs (which could destroy opposing ICBMs) *and* air defense against bombers, ballistic missile defense for cities, and civil defense. Seemingly "defensive" programs such as the Reagan Administration's ballistic missile defense ("Star Wars") program and parallel Soviet ballistic missile defense programs are in fact *offensive* under the inverted logic of a MAD world. See Posen and Van Evera, "Defense Policy and the Reagan Administration," pp. 24–25.

alerts would open and close larger windows, defensive military preparations would carry larger offensive implications, and smaller provocations could spur preemptive attack. Arms control would be harder to achieve, since secrecy would impede verification and treaties which met the security requirements of both sides would be harder to frame, which would circumscribe the ability of statesmen to escape this frightful world by agreement.

"Assured destruction" leaves much to be desired as a nuclear strategy, and the world of "mutual assured destruction" ("MAD") which it fosters leaves much to be desired as well. But 1914 warns that we tamper with MAD at our peril: any exit from MAD to a counterforce world would create a much more dangerous arrangement, whose outlines we glimpsed in the First World War.

Civil-Military Relations and the Cult of the Offensive, 1914 and 1984

Jack Snyder

Military technology should have made the European strategic balance in July 1914 a model of stability, but offensive military strategies defied those technological realities, trapping European statesmen in a war-causing spiral of insecurity and instability. As the Boer and Russo–Japanese Wars had foreshadowed and the Great War itself confirmed, prevailing weaponry and means of transport strongly favored the defender. Tactically, withering firepower gave a huge advantage to entrenched defenders; strategically, defenders operating on their own territory could use railroads to outmaneuver marching invaders. Despite these inexorable constraints, each of the major continental powers began the war with an offensive campaign. These war plans and the offensive doctrines behind them were in themselves an important and perhaps decisive cause of the war. Security, not conquest, was the principal criterion used by the designers of the plans, but their net effect was to reduce everyone's security and to convince at least some states that only preventive aggression could ensure their survival.

Even if the outbreak of war is taken as a given, the offensive plans must still be judged disasters. Each offensive failed to achieve its ambitious goals and, in doing so, created major disadvantages for the state that launched it. Germany's invasion of Belgium and France ensured that Britain would join the opposing coalition and implement a blockade. The miscarriage of France's ill-conceived frontal attack almost provided the margin of help that the Schlieffen Plan needed. Though the worst was averted by a last-minute railway maneuver, the Germans nonetheless occupied a key portion of France's industrial northeast, making a settlement based on the status quo ante impossible to negotiate. Meanwhile, in East Prussia the annihilation of an over-extended Russian invasion force squandered troops that might have

Robert Jervis, William McNeill, Cynthia Roberts, and Stephen Van Evera provided helpful comments on this paper, which draws heavily on the author's forthcoming book, *The Ideology of the Offensive: Military Decision Making and the Disasters of 1914* (Ithaca, N.Y.: Cornell University Press, 1984).

Jack Snyder is an Assistant Professor in the Political Science Department, Columbia University.

International Security, Summer 1984 (Vol. 9, No. 1) 0162-2889/84/010108-39 $02.50/1

been decisive if used to reinforce the undermanned advance into Austria. In each case, a defensive or more limited offensive strategy would have left the state in a more favorable strategic position.

None of these disasters was unpredictable or unpredicted. It was not only seers like Ivan Bloch who anticipated the stalemated positional warfare. General Staff strategists themselves, in their more lucid moments, foresaw these outcomes with astonishing accuracy. Schlieffen directed a war game in which he defeated his own plan with precisely the railway maneuver that Joffre employed to prevail on the Marne. In another German war game, which actually fell into Russian hands, Schlieffen used the advantage of railway mobility to defeat piecemeal the two prongs of a Russian advance around the Masurian Lakes—precisely the maneuver that led to the encirclement of Sazonov's Second Army at Tannenberg in August 1914. This is not to say that European war planners fully appreciated the overwhelming advantages of the defender; partly they underrated those advantages, partly they defied them. The point is that our own 20/20 hindsight is not qualitatively different from the understanding that was achievable by the historical protagonists.[1]

Why then were these self-defeating, war-causing strategies adopted? Although the particulars varied from country to country, in each case strategic policymaking was skewed by a pathological pattern of civil-military relations that allowed or encouraged the military to use wartime operational strategy to solve its institutional problems. When strategy went awry, it was because a penchant for offense helped the military organization to preserve its autonomy, prestige, and traditions, to simplify its institutional routines, or to resolve a dispute within the organization. As further discussion will show, it was not just a quirk of fate that offensive strategies served these functions. On balance, offense tends to suit the needs of military organizations better than defense does, and militaries normally exhibit at least a moderate preference for offensive strategies and doctrines for that reason. What was special about the period before World War I was that the state of civil-military relations in each of the major powers tended to exacerbate that normal offensive bias, either because the lack of civilian control allowed it to grow

---

1. Gerhard Ritter, *The Schlieffen Plan* (New York: Praeger, 1958), p. 60, note 34; A.A. Polivanov, *Voennoe delo*, No. 14 (1920), p. 421, quoted in Jack Snyder, *The Ideology of the Offensive: Military Decision Making and the Disasters of 1914* (Ithaca: Cornell University Press, 1984), chapter 7.

unchecked or because an abnormal degree of civil-military conflict heightened the need for a self-protective ideology.

In part, then, the "cult of the offensive" of 1914 reflected the endemic preference of military organizations for offensive strategies, but it also reflected particular circumstances that liberated or intensified that preference. The nature and timing of these catalytic circumstances, though all rooted in problems of civil-military relations, were different in each country. Indeed, if war had broken out as late as 1910, the Russian and French armies would both have fought quite defensively.[2]

Germany was the first European power to commit itself to a wildly overambitious offensive strategy, moving steadily in this direction from 1891 when Schlieffen became the Chief of the General Staff. The root of this pathology was the complete absence of civilian control over plans and doctrine, which provided no check on the natural tendency of mature military organizations to institutionalize and dogmatize doctrines that support the organizational goals of prestige, autonomy, and the elimination of novelty and uncertainty. Often, as in this case, it is offense that serves these interests best.[3]

France moved in 1911 from a cautious counteroffensive strategy towards the reckless frontal assault prescribed by the *offensive à outrance.* The roots of this doctrine can also be traced to a problem in civil-military relations. The French officer corps had always been wary of the Third Republic's inclination towards shorter and shorter terms of military service, which threatened the professional character and traditions of their organization. Touting the offense was a way to contain this threat, since everyone agreed that an army based on reservists and short-service conscripts would be good only for defense. The Dreyfus Affair and the radical military reforms that followed it heightened the officer corps' need for a self-protective ideology that would justify the essence and defend the autonomy of their organization. The extreme doctrine of the *offensive à outrance* served precisely this function, helping to discredit the defensive, reservist-based plans of the politicized

---

2. One reason that the war did not happen until 1914 was that Russian offensive power did not seriously threaten Germany until about that year. In this sense, the fact that all the powers had offensive strategies in the year the war broke out is to be explained more by their strategies' interactive consequences than by their common origins.
3. Snyder, *Ideology of the Offensive,* chapters 1, 4, and 5. I have profited greatly from the works of Barry Posen, *The Sources of Military Doctrine* (Ithaca: Cornell University Press, 1984), and Stephen Van Evera, "Causes of War" (Ph.D. dissertation, University of California, Berkeley, 1984), who advance similar arguments.

"republican" officers who ran the French military under civilian tutelage until the Agadir crisis of 1911. Given a freer rein in the harsher international climate, General Joffre and the Young Turks around him used the offensive doctrine to help justify a lengthening of the term of service and to reemphasize the value of a more highly professionalized army.[4]

Russia's drift towards increasingly overcommitted offensive plans between 1912 and 1914 was also abetted by the condition of civil-military relations. The problem in this case was the existence of two powerful veto groups within the military, one in the General Staff that favored an offensive against Germany and another centered on the Kiev military district that wanted to attack Austria. Forces were insufficient to carry out both missions, but there was no strong, centralized civilian authority who could or would enforce a rational priority commensurate with Russian means. Lacking firm civilian direction, the two military factions log-rolled the issue, each getting to implement its preferred offensive but with insufficient troops.[5]

It might be argued that these pathologies of civil-military relations are unique to the historical setting of this period. Civilians may have been ignorant of military affairs in a way that has been unequaled before or since. The transition in this period of the officer corps from an aristocratic caste to a specialized profession may have produced a uniquely unfavorable combination of the ill effects of both. Finally, social changes associated with rapid industrialization and urbanization may have provided a uniquely explosive setting for civil-military relations, as class conflicts reinforced civil-military conflicts.[6] Even if this is true, however, the same general patterns may persist but with lesser intensity, and understanding the circumstances that provoke more intense manifestations may help to forestall their recurrence.

Such a recurrence, whether intense or mild, is not a farfetched scenario. As in 1914, today's military technologies favor the defender of the status quo, but the superpowers are adopting offensive counterforce strategies in defiance of these technological constraints. Like machine guns and railroads, survivable nuclear weapons render trivial the marginal advantages to be gained by striking first. In the view of some, this stabilizing effect even neutralizes whatever first-strike advantages may exist at the conventional level, since the fear of uncontrollable escalation will restrain even the first

4. Snyder, *Ideology of the Offensive,* chapters 2 and 3. See also Samuel Williamson, *The Politics of Grand Strategy* (Cambridge: Harvard University Press, 1969).
5. Snyder, *Ideology of the Offensive,* chapters 6 and 7. See also A.M. Zaionchkovskii, *Podgotovka Rossii k imperialisticheskoi voine* (Moscow: Gosvoenizdat, 1926).
6. Van Evera, "Causes of War," chapter 7, explores these questions briefly.

steps in that direction. Since the would-be aggressor has the "last clear chance" to avoid disaster and normally cares less about the outcome than the defender does, mutual assured destruction works strongly for stability and the defense of the status quo. In this way, the absolute power to inflict punishment eases the security dilemma. All states possessing survivable second-strike forces can be simultaneously secure.[7]

Even those who are not entirely satisfied by the foregoing line of argument—and I include myself among them—must nevertheless admit the restraining effect that the irrevocable power to punish has had on international politics. Caveats aside, the prevailing military technology tends to work for stability, yet the strategic plans and doctrines of both superpowers have in important ways defied and undermined that basic reality. As in 1914, the danger today is that war will occur because of an erroneous belief that a disarming, offensive blow is feasible and necessary to ensure the attacker's security.

In order to understand the forces that are eroding the stability of the strategic balance in our own era, it may be helpful to reflect on the causes and consequences of the "cult of the offensive" of 1914. In proceeding towards this goal, I will discuss, first, how offensive strategies promoted war in 1914 and, second, why each of the major continental powers developed offensive military strategies. Germany will receive special attention because the Schlieffen Plan was the mainspring tightening the European security dilemma in 1914, because the lessons of the German experience can be more broadly generalized than those of the other cases, and because of the need to correct the widespread view that Germany's military strategy was determined by its revisionist diplomatic aims. After examining the domestic sources of military strategy in Germany, France, and Russia, I will discuss the effect of each state's policies on the civil-military relations and strategies of its neighbors. A concluding section will venture some possible applications of these findings to the study of contemporary Soviet military doctrine.

How Offense Promoted War

Conventional wisdom holds that World War I was caused in part by runaway offensive war plans, but historians and political scientists have been remark-

---

7. The best and most recent expression of this view is Robert Jervis, *The Illogic of American Nuclear Strategy* (Ithaca: Cornell University Press, 1984).

ably imprecise in reconstructing the logic of this process. Their vagueness has allowed critics of arms controllers' obsession with strategic instability to deny that the war resulted from "the reciprocal fear of surprise attack" or from any other by-product of offensive strategy.[8] Stephen Van Evera's contribution to this issue takes a major step towards identifying the manifold ways in which offensive strategies and doctrines promoted war in 1914. I would add only two points to his compelling argument. The first identifies some remaining puzzles about the perception of first-strike advantage in 1914; the second elaborates on Germany's incentive for preventive attack as the decisive way in which offensive military strategy led Europe towards war.

Van Evera cites statements and behavior indicating that European military and political decision-makers believed that the first army to mobilize and strike would gain a significant advantage. Fearing that their own preparations were lagging (or hoping to get a jump on the opponent), authorities in all of the countries felt pressed to take military measures that cut short the process of diplomacy, which might have converged on the solution of a "halt in Belgrade" if given more time. What is lacking in this story is a clear explanation of how the maximum gain or loss of two days could decisively affect the outcome of the campaign.

Planning documents suggest that no one believed that a two-day edge would allow a disarming surprise attack. Planners in all countries guarded against preemptive attacks on troops disembarking at railheads by concentrating their forces out of reach of such a blow. The only initial operation that depended on this kind of preemptive strike against unprepared forces was the German *coup de main* against the Belgian transport bottleneck of Liège. As the July crisis developed, the German General Staff was caused some anxiety by the progress of Belgian preparations to defend Liège, which jeopardized the smooth implementation of the Schlieffen Plan, but Moltke's attitude was not decisively influenced by this incentive to preempt.[9] In any event, it was Russia that mobilized first, and there is little to suggest that preemption was decisive in this case either. Prewar planning documents and

8. Even the usually crystal-clear Thomas Schelling is a bit murky on this point. See his *Arms and Influence* (New Haven: Yale University Press, 1966), pp. 221–225. For a critic, see Stephen Peter Rosen, "Nuclear Arms and Strategic Defense," *Washington Quarterly*, Vol. 4, No. 2 (Spring 1981), pp. 83–84.
9. Ulrich Trumpener, "War Premeditated? German Intelligence Operations in July 1914," *Central European History*, Vol. 9, No. 1 (March 1976), p. 80.

staff exercises show that the Russians worried about being preempted, but took sufficient precautions against it. They also indicate that preemption was not particularly feared if Austria was embroiled in the Balkans—precisely the conditions that obtained in July 1914. On the offensive side, however, the incentive to strike first might have been an important factor. Van Evera points out that the difference between the best case (mobilizing first) and the worst case (mobilizing second) was probably a net gain of four days (two gained plus two not lost). Given the Russians' aim of putting pressure on Germany's rear before the campaign in France was decided, four days was not a negligible consideration. To save just two days, the Russians were willing to begin their advance without waiting for the formation of their supply echelons. Thus, time pressure imposed by military exigencies may explain the haste of the crucial Russian mobilization. It should be stressed, however, that it was neither "the reciprocal fear of surprise attack" nor the chance of preempting the opponent's unalerted forces that produced this pressure. Rather, it was the desire to close Germany's window of opportunity against France that gave Russia an incentive to strike first.[10]

A second elaboration of Van Evera's argument, which will be crucial for understanding the following sections of this paper, is that offensive plans not only reflected the belief that states are vulnerable and conquest is easy; they actually caused the states adopting them to *be* vulnerable and consequently fearful. Even the Fischer school, which emphasizes Germany's "grasping for 'World Power'" as the primary cause of the war, admits that Germany's decision to provoke a conflict in 1914 was also due to the huge Russian army increases then in progress, which would have left Germany at Russia's mercy upon their completion in 1917.[11] This impending vulnerabil-

---

10. Russia, 10-i otdel General'nogo shtaba RKKA, *Vostochnoprusskaia operatsiia: sbornik dokumentov* (Moscow: Gosvoenizdat, 1939), especially p. 62, which reproduces a Russian General Staff intelligence estimate dated March 1, 1914. Van Evera's quotations suggest that decision-makers in all countries exhibited more concern about being preempted than seems warranted by actual circumstances. One explanation may be that the military oversold this danger as a way of guarding against the risk of excessive civilian foot-dragging, which was clearly a concern among the French military, at least. Another possibility is that there was a disconnect between the operational level of analysis, where it was obvious that no one could disrupt his opponent's concentration, and the more abstract level of doctrine, where the intangible benefits of "seizing the initiative" were nonetheless considered important. See Snyder, *Ideology of the Offensive*, chapters 2 and 3.

11. The Germans saw the planned 40 percent increase in the size of the Russian standing army as a threat to Germany's physical survival, not just a barrier foreclosing opportunities to expand. This is expressed most clearly in the fear that the power shift would allow Russia to force a

ity, though real enough, was largely a function of the Schlieffen Plan, which had to strip the eastern front in order to amass the forces needed to deal with the strategic conundrums and additional opponents created by the march through Belgium. If the Germans had used a positional defense on the short Franco–German border to achieve economies of force, they could have handled even the enlarged Russian contingents planned for 1917.[12]

In these ways, offensive strategies helped to cause the war and ensured that, when war occurred, it would be a world war. Prevailing technologies should have made the world of 1914 an arms controllers' dream; instead, military planners created a nightmare of strategic instability.

Germany: Uncontrolled Military or Militarized Civilians?

The offensive character of German war planning in the years before World War I was primarily an expression of the professional interests and outlook of the General Staff. Civilian foreign policy aims and attitudes about international politics were at most a permissive cause of the Schlieffen Plan. On balance, the General Staff's all-or-nothing war plan was more a hindrance than a help in implementing the diplomats' strategy of brinkmanship. The reason that the military was allowed to indulge its strategic preferences was not so much that the civilians agreed with them; rather, it was because war planning was considered to be within the autonomous purview of the General Staff. Military preferences were never decisive on questions of the use of force, however, since this was not considered their legitimate sphere. But indirectly, war plans trapped the diplomats by handing them a blunt instrument suitable for massive preventive war, but ill-designed for controlled coercion. The military's unchecked preference for an unlimited offensive strategy and the mismatch between German military and diplomatic strategy were important causes of strategic instability rooted in the problem of civil-military relations. This section will trace those roots and point out some implications relevant to contemporary questions.

The Schlieffen Plan embodied all of the desiderata commonly found in field manuals and treatises on strategy written by military officers: it was an

---

revision of the status quo in the Balkans, leading to Austria's collapse. See especially Fritz Fischer, *War of Illusions: German Policies from 1911 to 1914* (New York: W.W. Norton, 1975; German edition 1969), pp. 377–379, 427.

12. This is argued in Snyder, *Ideology of the Offensive*, chapter 4.

offensive campaign, designed to seize the initiative, to exploit fleeting opportunities, and to achieve a decisive victory by the rapid annihilation of the opponents' military forces. War was to be an "instrument of politics," not in the sense that political ends would restrain and shape military means, but along lines that the General Staff found more congenial: war would solve the tangle of political problems that the diplomats could not solve for themselves. "The complete defeat of the enemy always serves politics," argued General Colmar von der Goltz in his influential book, *The Nation in Arms*. "Observance of this principle not only grants the greatest measure of freedom in the political sphere but also gives widest scope to the proper use of resources in war."[13]

To do this, Schlieffen sought to capitalize on the relatively slow mobilization of the Russian army, which could not bring its full weight to bear until the second month of the campaign. Schlieffen reasoned that he had to use this "window of opportunity" to decisively alter the balance of forces in Germany's favor. Drawing on precedents provided by Moltke's campaigns of 1866 and 1870 as well as his later plans for a two-front war, Schlieffen saw that a rapid decision could be achieved only by deploying the bulk of the German army on one front in order to carry out a grandiose encirclement maneuver. France had to be the first victim, because the Russians might spoil the encirclement by retreating into their vast spaces. With Paris at risk, the French would have to stand and fight. By 1897, Schlieffen had concluded that this scheme could not succeed without traversing Belgium, since the Franco–German frontier in Alsace–Lorraine was too narrow and too easily defended to permit a decisive maneuver. In the mature conception of 1905, most of the German army (including some units that did not yet exist) would march for three or four weeks through Belgium and northern France, encircling and destroying the French army, and then board trains for the eastern front to reinforce the few divisions left to cover East Prussia.

Even Schlieffen was aware that his plan was "an enterprise for which we are too weak."[14] He and his successor, the younger Moltke, understood most of the pitfalls of this maneuver quite well: the gratuitous provocation of new enemies, the logistical nightmares, the possibility of a rapid French rede-

13. Gerhard Ritter, *The Sword and the Scepter: The Problem of Militarism in Germany* (Coral Gables: University of Miami Press, 1969; German edition 1954), Vol. 1, p. 196, citing *Das Volk in Waffen* (5th ed., 1889), p. 129.
14. Ritter, *Schlieffen Plan*, p. 66.

ployment to nullify the German flank maneuver, the numerical insufficiency of the Germany army, the tendency of the attacker's strength to wane with every step forward and the defender's to grow, and the lack of time to finish with France before Russia would attack. The General Staff clung to this plan not because they were blind to its faults, but because they thought all the alternatives were worse. To mollify Austria in 1912, they went through the motions of gaming out a mirror-image of the Schlieffen Plan pointed towards the east, concluding that the French would defeat the weak forces left in the Rhineland long before a decision could be reached against Russia.[15] What the General Staff refused to consider seriously after 1890 was the possibility of an equal division of their forces between west and east, allowing a stable defensive against France and a limited offensive with Austria against Russia. (This was the combination that Germany used successfully in 1915 and that the elder Moltke had resigned himself to in the 1890s.)

Around the turn of the century, the General Staff played some war games based on a defensive in the west. These led to the embarrassing conclusion that the French would have great difficulty overwhelming even a modest defensive force. In future years, when games with this premise were played, the German defenders were allotted fewer forces, while Belgians and Dutch were arbitrarily added to the attacking force. Stacking the deck against the defensive appeared not only in war-gaming but also in Schlieffen's abstract expostulations of doctrine. Even some German critics caught him applying a double standard, arbitrarily granting the attacker advantages in mobility, whereas the reality should have been quite the opposite.[16]

In short, German war planning, especially after 1890, showed a strong bias in favor of offensive schemes for decisive victory and against defensive or more limited offensive schemes, even though the latter had a greater prospect of success. This bias cannot be explained away by the argument that Germany would have been at an economic disadvantage in a long war against Russia and hence had to gamble everything on a quick victory. As the actual war showed, this was untrue. More important, Schlieffen hit upon economic rationalizations for his war plan only after it had already been in place for years. Moreover, he actively discouraged serious analysis of wartime economics, deciding *a priori* that the only good war was a short war and that

---

15. Louis Garros, "Préludes aux invasions de la Belgique," *Revue historique de l'armée* (March 1949), pp. 37–38; French archival documents cited in Snyder, *Ideology of the Offensive*, chapter 4.
16. Friedrich von Bernhardi, *On War of Today* (London: Rees, 1912), Vol. 1, p. 44.

the only way to end a war quickly was to disarm the opponent decisively.[17] These conclusions were not in themselves unreasonable, but Schlieffen reached them before he did his analysis and then arranged the evidence in order to justify his preferred strategy.

The explanation for the General Staff's bias in favor of offensive strategy is rooted in the organizational interests and parochial outlook of the professional military. The Germans' pursuit of a strategy for a short, offensive, decisive war despite its operational infeasibility is simply an extreme case of an endemic bias of military organizations. Militaries do not always exhibit a blind preference for the offensive, of course. The lessons of 1914–1918 had a tempering effect on the offensive inclinations of European militaries, for example.[18] Still, exceptions and questionable cases notwithstanding, initial research indicates that militaries habitually prefer offensive strategies, even though everyone from Clausewitz to Trevor Dupuy has proved that the defender enjoys a net operational advantage.[19]

EXPLAINING THE OFFENSIVE BIAS

Several explanations for this offensive bias have been advanced. A number of them are consistent with the evidence provided by the German case. A particularly important explanation stems from the division of labor and the narrow focus of attention that necessarily follows from it. The professional training and duties of the soldier force him to focus on threats to his state's security and on the conflictual side of international relations. Necessarily preoccupied with the prospect of armed conflict, he sees war as a pervasive aspect of international life. Focusing on the role of military means in ensuring the security of the state, he forgets that other means can also be used towards that end. For these reasons, the military professional tends to hold a simplified, zero-sum view of international politics and the nature of war, in which wars are seen as difficult to avoid and almost impossible to limit.

---

17. Lothar Burchardt, *Friedenswirtschaft und Kriegsvorsorge: Deutschlands wirtschaftliche Rüstungsbestrebungen vor 1914* (Boppard am Rhein: Boldt, 1968), pp. 15, 163–164.
18. However, this effect should not be overdrawn. Barry Posen, *Sources of Military Doctrine,* has recently demonstrated that the French collapse in 1940 was due not to a Maginot Line mentality but to the overcommitment of forces to the offensive campaign in Belgium.
19. Possible biases in civilian views on offense and defense have not been studied systematically. For Trevor Dupuy's attempts to analyze quantitatively offensive and defensive operations in World War II, see his *Numbers, Predictions and War* (New York: Bobbs-Merrill, 1979), chapter 7, and other publications of his "HERO" project.

When the hostility of others is taken for granted, prudential calculations are slanted in favor of preventive wars and preemptive strikes. Indeed, as German military officers were fond of arguing, the proper role of diplomacy in a Hobbesian world is to create favorable conditions for launching preventive war. A preventive grand strategy requires an offensive operational doctrine. Defensive plans and doctrines will be considered only after all conceivable offensive schemes have been decisively discredited. Under uncertainty, such discrediting will be difficult, so offensive plans and doctrines will frequently be adopted even if offense is not easier than defense in the operational sense.

The assumption of extreme hostility also favors the notion that decisive, offensive operations are always needed to end wars. If the conflict of interest between the parties is seen as limited, then a decisive victory may not be needed to end the fighting on mutually acceptable terms. In fact, denying the opponent his objectives by means of a successful defense may suffice. However, when the opponent is believed to be extremely hostile, disarming him completely may seem to be the only way to induce him to break off his attacks. For this reason, offensive doctrines and plans are needed, even if defense is easier operationally.

Kenneth Waltz argues that states are socialized to the implications of international anarchy.[20] Because of their professional preoccupations military professionals become "oversocialized." Seeing war more likely than it really is, they increase its likelihood by adopting offensive plans and buying offensive forces. In this way, the perception that war is inevitable becomes a self-fulfilling prophecy.

A second explanation emphasizes the need of large, complex organizations to operate in a predictable, structured environment. Organizations like to work according to a plan that ties together the standard operating procedures of all the subunits into a prepackaged script. So that they can stick to this script at all costs, organizations try to dominate their environment rather than react to it. Reacting to unpredictable circumstances means throwing out the plan, improvising, and perhaps even deviating from standard operating procedures. As Barry Posen points out, "taking the offensive, exercising the initiative, is a way of structuring the battle."[21] Defense, in contrast, is more reactive, less structured, and harder to plan. Van Evera argues that the

20. Kenneth N. Waltz, *Theory of International Politics* (Reading, Mass.: Addison-Wesley, 1979).
21. Posen, *Sources of Military Doctrine,* chapter 2.

military will prefer a task that is easier to plan even if it is more difficult to execute successfully.[22] In Russia, for example, regional staffs complained that the General Staff's defensive war plan of 1910 left their own local planning problem too unstructured. They clamored for an offensive plan with specified lines of advance, and in 1912 they got it.[23]

The German military's bias for the offensive may have derived in part from this desire to structure the environment, but evidence on this point is mixed. The elder Moltke developed clockwork mobilization and rail transport plans leading to offensive operations, but he scoffed at the idea that a campaign plan could be mapped out step-by-step from the initial deployment through to the crowning encirclement battle. For him, strategy remained "a system of *ad hoc* expedients . . . , the development of an original idea in accordance with continually changing circumstances."[24] This attitude may help to explain his willingness to entertain defensive alternatives when his preferred offensive schemes began to look too unpromising. The Schlieffen Plan, in contrast, was a caricature of the link between rigid planning and an unvarying commitment to the offensive. Even here, however, there is some evidence that fits poorly with the hypothesis that militaries prefer offense because it allows them to fight according to their plans and standard operating procedures. Wilhelm Groener, the General Staff officer in charge of working out the logistical preparations for the Schlieffen Plan, recognized full well that the taut, ambitious nature of the plan would make it impossible to adhere to normal, methodical supply procedures. Among officers responsible for logistics, "the feeling of responsibility must be so great that in difficult circumstances people free themselves from procedural hindrances and take the responsibility for acting in accordance with common sense."[25] Nonetheless, it is difficult to ignore the argument ubiquitously advanced by European military writers that defense leads to uncertainty, confusion, passivity, and incoherent action, whereas offense focuses the efforts of the army and the mind of the commander on a single, unwavering goal. Even when they understood the uncertainties and improvisations required by offensive operations, as Groener did, they may still have feared the uncertainties of the defensive more. An offensive plan at least gives the illusion of certainty.

---

22. Van Evera, "Causes of War," chapter 7.
23. Zaionchkovskii, *Podgotovka Rossii k imperialisticheskoi voine*, pp. 244, 277.
24. Quoted by Hajo Holborn, "Moltke and Schlieffen," in Edward M. Earle, ed., *Makers of Modern Strategy* (Princeton: Princeton University Press, 1971), p. 180.
25. Papers of Wilhelm Groener, U.S. National Archives, roll 18, piece 168, p. 5.

Another possibility, however, is that this argument for the offensive was used to justify a doctrine that was preferred primarily on other grounds. French military publicists invoked such reasoning more frequently, for example, during periods of greater threat to traditional military institutions.[26]

Other explanations for the offensive bias are rooted even more directly in the parochial interests of the military, including the autonomy, prestige, size, and wealth of the organization.[27] The German case shows the function of the offensive strategy as a means towards the goal of operational autonomy. The elder Moltke succinctly stated the universal wish of military commanders: "The politician should fall silent the moment that mobilization begins."[28] This is least likely to happen in the case of limited or defensive wars, where the whole point of fighting is to negotiate a diplomatic solution. Political considerations—and hence politicians—have to figure in operational decisions. The operational autonomy of the military is most likely to be allowed when the operational goal is to disarm the adversary quickly and decisively by offensive means. For this reason, the military will seek to force doctrine and planning into this mold.

The prestige, self-image, and material health of military institutions will prosper if the military can convince civilians and themselves that wars can be short, decisive, and socially beneficial. One of the attractions of decisive, offensive strategies is that they hold out the promise of a demonstrable return on the nation's investment in military capability. Von der Goltz, for example, pushed the view that "modern wars have become the nation's way of doing business"—a perspective that made sense only if wars were short, cheap, and hence offensive.[29] The German people were relatively easy to convince of this, because of the powerful example provided by the short, offensive, nation-building wars of 1866 and 1870, which cut through political fetters and turned the officer corps into demigods. This historical backdrop gave the General Staff a mantel of unquestioned authority and legitimacy in operational questions; it also gave them a reputation to live up to. Later, when technological and strategic circumstances challenged the viability of their

26. See the argument in Snyder, *Ideology of the Offensive,* chapter 3, citing especially Georges Gilbert, *Essais de critique militaire* (Paris: Librairie de la Nouvelle Revue, 1890), pp. 43, 47–48.
27. Posen and Van Evera, in analyzing organizational interests in this way, have drawn on the categories laid out by Morton Halperin, *Bureaucratic Politics and Foreign Policy* (Washington: Brookings, 1974), chapter 3.
28. Quoted by Bernard Brodie, *War and Politics* (New York: Macmillan, 1973), p. 11.
29. Quoted by Van Evera from Ferdinand Foch, *The Principles of War* (New York: Fly, 1918), p. 37.

formula for a short, victorious war, General Staff officers like Schlieffen found it difficult to part with the offensive strategic formulae that had served their state and organization so effectively. As Posen puts it, offense makes soldiers "specialists in victory," defense makes them "specialists in attrition," and in our own era mutual assured destruction makes them "specialists in slaughter."[30]

THE EVOLUTION OF GERMAN WAR PLANNING

The foregoing arguments could, for the most part, explain the offensive bias of the military in many countries and in many eras. What remains to be explained is why this offensive bias became so dogmatic and extreme in Germany before 1914. The evolution of the General Staff's strategic thinking from 1870 to 1914 suggests that a tendency towards doctrinal dogmatism and extremism may be inherent in mature military organizations that develop under conditions of near-absolute autonomy in doctrinal questions. This evolution, which occurred in three stages, may be typical of the maturation of uncontrolled, self-evaluating organizations and consequently may highlight the conditions in which doctrinal extremism might recur in our own era.[31]

The first stage was dominated by the elder Moltke, who established the basis tenets of the organizational ideology of the German General Staff. These were the inevitability and productive nature of war, the indispensability of preventive war, and the need for an operational strategy that could provide rapid, decisive victories. Moltke was the creator, not a captive of his doctrines and did not implement them in the manner of a narrow technician. He was willing to think in political terms and to make his opinion heard in political matters. This practice had its good and bad sides. On one hand, it allowed him to consider war plans that gave diplomacy some role in ending the war; on the other, it spurred him to lobby for preventive war against France in 1868 and against Russia in 1887. Moltke thought he understood what international politics was all about, but he understood it in a military way. In judging the opportune moment for war, Moltke looked exclusively at military factors, whereas Bismarck focused primarily on preparing domestic and foreign opinion for the conflict.[32]

---

30. Posen, *Sources of Military Doctrine*.
31. Van Evera uses the concept of the self-evaluating organization, drawing on the work of James Q. Wilson.
32. Ritter, *Sword and Scepter*, Vol. 1, pp. 217–218, 245.

Schlieffen, the key figure in the second stage of the General Staff's development, was much more of a technocrat than Moltke. Not a founder, he was a systematizer and routinizer. Schlieffen dogmatized Moltke's strategic precepts in a way that served the mature institution's need for a simple, standardized doctrine to facilitate the training of young officers and the operational planning of the General Staff. In implementing this more dogmatic doctrine, Schlieffen and his colleagues lacked Moltke's ability to criticize fundamental assumptions and tailor doctrine to variations in circumstances. Thus, Moltke observed the defender's increasing advantages and decided reluctantly that the day of the rapid, decisive victory was probably gone, anticipating that "two armies prepared for battle will stand opposite each other, neither wishing to begin battle."[33] Schlieffen witnessed even further developments in this direction in the Russo–Japanese War, but concluded only that the attacker had to redouble his efforts. "The armament of the army has changed," he recognized, "but the fundamental laws of combat remain the same, and one of these laws is that one cannot defeat the enemy without attacking."[34]

Seeing himself as primarily a technician, Schlieffen gave political considerations a lesser place in his work than had Moltke. Again, this had both good and bad consequences. On one hand, Schlieffen never lobbied for preventive war in the way Moltke and Waldersee had, thinking such decisions were not his to make. When asked, of course, he was not reluctant to tell the political authorities that the time was propitious, as he did in 1905. On the other hand, Schlieffen had a more zero-sum, apolitical view of the conduct of warfare than did the elder Moltke. Consequently, his war plans excluded any notion of political limitations on the conduct of war or diplomatic means to end it.[35]

Contrasting the problems of civilian control of the military in stages one and two, we see that the founders' generation, being more "political," chal-

---

33. Helmuth von Moltke, *Die Deutschen Aufmarschpläne, 1871–1890,* Ferdinand von Schmerfeld, ed. (Berlin: Mittler, 1929), p. 122ff.

34. The quotation is from an 1893 comment on an operational exercise, quoted by O. von Zoellner, "Schlieffens Vermächtnis," *Militärwissenschaftliche Rundschau* (Sonderheft, 1938), p. 18, but identical sentiments are expressed in Schlieffen's "Krieg in der Gegenwart," *Deutsche Revue* (1909).

35. Brodie, *War and Politics,* p. 58, reports a perhaps apocryphal statement by Schlieffen that if his plan failed to achieve decisive results, then Germany should negotiate an end to the war. Even if he did say this, the possibility of negotiations had no effect on his war planning, in contrast to that of the elder Moltke.

lenges the political elite on questions of the use of force, but as if in compensation, is more capable of self-evaluation and self-control in its war planning. The technocratic generation, however, is less assertive politically but also less capable of exercising political judgment in its own work. The founders' assertiveness is the more dramatic challenge to political control, but as the German case shows, Bismarck was able to turn back the military's direct lobbying for preventive war, which was outside of the military's legitimate purview even by the Second Reich's skewed standards of civil-military relations. Much more damaging in the long run was Schlieffen's unobtrusive militarism, which created the conditions for a preventive war much more surely than Moltke's overt efforts did.

A third stage, which was just developing on the eve of World War I, combined the worst features of the two previous periods. Exemplary figures in this final stage were Erich Ludendorff and Wilhelm Groener, products of a thoroughgoing socialization to the organizational ideology of the German General Staff. Groener, describing his own war college training, makes it clear that not only operational principles but also a militaristic philosophy of life were standard fare in the school's curriculum. These future functionaries and leaders of the General Staff were getting an intensive course in the same kind of propaganda that the Army and Navy Leagues were providing the general public. They came out of this training believing in the philosophy of total war, demanding army increases that their elders were reluctant to pursue and fearing that "weaklings" like Bethmann Hollweg would throw away the army's glorious victories.[36]

An organizational explanation for this third stage would point to the self-amplifying effects of the organizational ideology in a mature, self-evaluating unit. An alternative explanation also seems plausible, however. Geoff Eley, in his study of right-wing radical nationalism in Wilhelmine Germany, argues that emerging counterelites used national populist causes and institutions like the Navy and Army Leagues as weapons aimed at the political monopoly retained by the more cautious traditional elite, who were vulnerable to criticism on jingoistic issues.[37] This pattern fits the cases of Groener and Ludendorff, who were middle-class officers seeking the final transformation of the

36. Helmut Haeussler, *General William Groener and the Imperial German Army* (Madison: The State Historical Society of Wisconsin, 1962), p. 72.
37. Geoff Eley, *Reshaping the German Right: Radical Nationalism and Political Change after Bismarck* (New Haven: Yale University Press, 1980).

old Prussian army into a mass organ of total war, which would provide upward mobility for their own kind. German War Ministers, speaking for conservative elements in the army and the state, had traditionally resisted large increases in the size of the army, which would bring more bourgeois officers into the mess and working-class soldiers into the ranks; it would also cost so much that the Junkers' privileged tax status would be brought into question. This alternative explanation makes it difficult to know whether organizational ideologies really tend toward self-amplification or whether extremist variants only occur from some particular motivation, as the French case suggests.

THE MISMATCH BETWEEN MILITARY STRATEGY AND DIPLOMACY

It is sometimes thought that Germany required an unlimited, offensive military strategy because German civilian elites were hell-bent on overturning the continental balance of power as a first step in their drive for "World Power." In this view, the Schlieffen Plan was simply the tool needed to achieve this high-risk, high-payoff goal, around which a national consensus of both military and civilian elites had formed.[38] There are several problems with this view. The first is that the civilians made virtually no input into the strategic planning process. Contrary to the unsupported assertions of some historians, the shift from Moltke's plan for a limited offensive against Russia to Schlieffen's plan for a more decisive blow aimed at France had nothing to do with the fall of Bismarck or the "New Course" in foreign policy. Rather, Schlieffen saw it as a technical change, stemming from an improved Russian ability to defend their forward theater in Poland. Nor was Schlieffen chosen to head the General Staff because of the strategy he preferred. Schlieffen had simply been the next in line as deputy chief under Waldersee, who was fired primarily because he dared to criticize the Kaiser's tactical decisions in a mock battle.[39] Later, when Reich Chancellor von Bülow learned of Schlieffen's intention to violate Belgian neutrality, his reaction was: "if the Chief of Staff, especially a strategic authority such as Schlieffen, believes such a measure to be necessary, then it is the obligation of diplomacy to adjust to it and prepare for it in every possible way."[40] In 1912 Foreign Secretary von

---

38. See, for example, L.L. Farrar, Jr., *Arrogance and Anxiety* (Iowa City: University of Iowa Press, 1981), pp. 23–24.
39. Ritter, *Schlieffen Plan,* pp. 17–37; Norman Rich and M.H. Fisher, eds., *The Holstein Papers* (Cambridge: Cambridge University Press, 1963), Vol. 3, pp. 347, note 1, and 352–353.
40. Ritter, *Schlieffen Plan,* pp. 91–92.

Jagow urged a reevaluation of the need to cross Belgian territory, but a memo from the younger Moltke ended the matter.[41] In short, the civilians knew what Schlieffen was planning to do, but they were relatively passive bystanders in part because military strategy was not in their sphere of competence and legitimate authority, and perhaps also because they were quite happy with the notion that the war could be won quickly and decisively. This optimism alleviated their fear that a long war would mean the destruction of existing social and economic institutions, no matter who won it. The decisive victory promised by the Schlieffen Plan may have also appealed to civilian elites concerned about the need for spectacular successes as a payoff for the masses' enthusiastic participation in the war. Trying to justify the initial war plan from the retrospective vantage point of 1919, Bethmann Hollweg argued that "offense in the East and defense in the West would have implied that we expected at best a draw. With such a slogan no army and no nation could be led into a struggle for their existence."[42] Still, this is a long way from the totally unfounded notion that Holstein and Schlieffen cooked up the Schlieffen Plan expressly for the purpose of bullying France over the Morocco issue and preparing the way for "Welt Politik."[43] The Schlieffen Plan had some appeal for German civilian elites, but the diplomats may have had serious reservations about it, as the Jagow episode suggests. Mostly, the civilians passively accepted whatever operational plan the military deemed necessary.

If German diplomats had devised a military strategy on their own, it is by no means certain that they would have come up with anything like the Schlieffen Plan. This all-or-nothing operational scheme fit poorly with the diplomatic strategy of expansion by means of brinkmanship and controlled, coercive pressure, which they pursued until 1914. In 1905, for example, it is clear that Bülow, Holstein, and Wilhelm II had no inclination to risk a world war over the question of Morocco.

"The originators of *Weltpolitik* looked forward to a series of small-scale, marginal foreign policy successes," says historian David Kaiser, "not to a major war."[44] Self-deterred by the unlimited character of the Schlieffen Plan,

41. Fischer, *War of Illusions*, p. 390.
42. Konrad Jarausch, *The Enigmatic Chancellor* (New Haven: Yale University Press, 1973), p. 195.
43. This is implied by Martin Kitchen, *The German Officer Corps* (Oxford: Oxford University Press, 1968), p. 104, and Imanuel Geiss, *German Foreign Policy, 1871–1914* (London: Routledge & Kegan Paul, 1976), pp. 101–103.
44. David E. Kaiser, "Germany and the Origins of the First World War," *Journal of Modern History*, Vol. 55, No. 3 (September 1983), p. 448.

they had few military tools that they could use to demonstrate resolve in a competition in risk-taking. The navy offered a means for the limited, demonstrative use of force, namely the dispatch of the gunboat *Panther* to the Moroccan port of Agadir, but the army was an inflexible tool. At one point in the crisis, Schlieffen told Bülow that the French were calling up reservists on the frontier. If this continued, Germany would have to respond, setting off a process that the Germans feared would be uncontrollable.[45] Thus, the German military posture and war plan served mainly to deter the German diplomats, who did not want a major war even though Schlieffen told them the time was favorable. They needed limited options, suitable for coercive diplomacy, not unlimited options, suitable for preventive war. With the Schlieffen Plan, they could not even respond to the opponent's precautionary moves without setting off a landslide toward total war.

This mismatch between military and diplomatic strategy dogged German policy down through 1914. Bethmann Hollweg described his strategy in 1912 as one of controlled coercion, sometimes asserting German demands, sometimes lulling and mollifying opponents to control the risk of war. "On all fronts we must drive forward quietly and patiently," he explained, "without having to risk our existence."[46] Bethmann's personal secretary, Kurt Riezler, explained this strategy of calculated risk in a 1914 volume, *Grundzüge der Weltpolitik.* A kind of cross between Thomas Schelling and Norman Angell, Riezler explained that wars were too costly to actually fight in the modern, interdependent, capitalist world. Nonetheless, states can still use the threat of war to gain unilateral advantages, forcing the opponent to calculate whether costs, benefits, and the probability of success warrant resorting to force. His calculations can be affected in several ways. Arms-racing can be used, *á la* Samuel Huntington, as a substitute for war—that is, a bloodless way to show the opponent that he would surely lose if it came to a fight. Brinkmanship and bluffing can be used to demonstrate resolve; *faits accomplis* and salami tactics can be used to shift the onus for starting the undesired war onto the opponent. But, Riezler warns, this strategy will not work if one is greedy and impatient to overturn the balance of power. Opponents will fight if they sense that their vital interests are at stake. Consequently, "victory

---

45. Holstein to Radolin, June 28, 1905, in *Holstein Papers,* Vol. 4, p. 347.
46. Jarausch, *Enigmatic Chancellor,* pp. 110–111.

belongs to the steady, tenacious, and gradual achievement of small successes . . . without provocation."[47]

Although this may have been a fair approximation of Bethmann's thinking in 1912, the theory of the calculated risk had undergone a major transformation by July 1914. By that time, Bethmann wanted a major diplomatic or military victory and was willing to risk a continental war—perhaps even a world war—to achieve it. *Fait accompli* and onus-shifting were still part of the strategy, but with a goal of keeping Britain out of the war and gaining the support of German socialists, not with a goal of avoiding war altogether.

The Schlieffen Plan played an important role in the transformation of Bethmann's strategy and in its failure to keep Britain neutral in the July crisis. Riezler's diary shows Bethmann's obsession in July 1914 with Germany's need for a dramatic victory to forestall the impending period of vulnerability that the Russian army increases and the possible collapse of Austria–Hungary would bring on.[48] As I argued earlier, the Schlieffen Plan only increased Germany's vulnerability to the Russian buildup, stripping the eastern front and squandering forces in the vain attempt to knock France out of the war. In this sense, it was the Schlieffen Plan that led Bethmann to transform the calculated-risk theory from a cautious tool of coercive diplomacy into a blind hope of gaining a major victory without incurring an unwanted world war.

Just as the Schlieffen Plan made trouble for Bethmann's diplomacy, so too German brinkmanship made trouble for the Schlieffen Plan. The Russian army increases, provoked by German belligerence in the 1909 Bosnian crisis and Austrian coercion of the Serbs in 1912, made the German war plan untenable.[49] The arms-racing produced by this aggressive diplomacy was not a "substitute for war"; rather, it created a window of vulnerability that helped to cause the war. Thus, Riezler (and Bethmann) failed to consider how easily a diplomatic strategy of calculated brinkmanship could set off a chain of uncontrollable consequences in a world of military instability.

---

47. Andreas Hillgruber, *Germany and the Two World Wars* (Cambridge: Harvard University Press, 1981), pp. 22–24; J.J. Ruedorffer (pseud. for Kurt Riezler), *Grundzüge der Weltpolitik in der Gegenwart* (Berlin: Deutsche Verlags-Anstalt, 1914), especially pp. 214–232; quotation from Jarausch, *Enigmatic Chancellor*, pp. 143–144.
48. Jarausch, *Enigmatic Chancellor*, p. 157.
49. P.A. Zhilin, "Bol'shaia programma po usileniiu russkoi armii," *Voenno-istoricheskii zhurnal*, No. 7 (July 1974), pp. 90–97, shows the connection between the 1913 increases and the Balkan crisis of 1912. He also shows that this project, with its emphasis on increasing the standing army and providing rail lines to speed its concentration, was directly connected to the offensive character of Russia's increasingly overcommitted, standing-start, short-war campaign plan.

Even the transformed version of the calculated-risk theory, implemented in July 1914, was ill-served by the Schlieffen Plan. If Bethmann had had eastern-oriented or otherwise limited military options, all sorts of possibilities would have been available for defending Austria, bloodying the Russians, driving a wedge between Paris and St. Petersburg, and keeping Britain neutral. In contrast, the Schlieffen Plan cut short any chance for coercive diplomacy and ensured that Britain would fight. In short, under Bethmann as well as Bülow, the Schlieffen Plan was hardly an appropriate tool underwriting the brinkmanship and expansionist aims of the civilian elite. Rather, the plan was the product of military organizational interests and misconceptions that reduced international politics to a series of preventive wars. The consequences of the all-or-nothing war plan were, first, to reduce the coercive bargaining leverage available to German diplomats, and second, to ensnare German diplomacy in a security dilemma that forced the abandonment of the strategy of controlled risks. Devised by military officers who wanted a tool appropriate for preventive war, the Schlieffen Plan trapped Germany in a situation where preventive war seemed like the only safe option.

In summary, three generalizations emerge from the German case. First, military organizations tend to exhibit a bias in favor of offensive strategies, which promote organizational prestige and autonomy, facilitate planning and adherence to standard operating procedures, and follow logically from the officer corps' zero-sum view of international politics. Second, this bias will be particularly extreme in mature organizations which have developed institutional ideologies and operational doctrines with little civilian oversight. Finally, the destabilizing consequences of an inflexible, offensive military strategy are compounded when it is mismatched with a diplomatic strategy based on the assumption that risks can be calculated and controlled through the skillful fine-tuning of threats.

France: Civil-Military Truce and Conflict

France before the Dreyfus Affair exemplifies the healthiest pattern of civil-military relations among the European states, but after Dreyfus, the most destructive. In the former period civilian defense experts who understood and respected the military contained the latent conflict between the professional army and republican politicians by striking a bargain that satisfied the main concerns of both sides. In this setting, the use of operational doctrine as a weapon of institutional defense was minimal, so plans and doctrine were a moderate combination of offense and defense. After the Dreyfus watershed, the truce broke. Politicians set out to "republicanize" the army, and the officer corps responded by developing the doctrine of *offensive à outrance*, which helped to reverse the slide towards a military system based overwhelmingly on reservists and capable only of defensive operations.[50]

The French army had always coexisted uneasily with the Third Republic. Especially in the early years, most officers were Bonapartist or monarchist in their political sentiments, and Radical politicians somewhat unjustifiably feared a military coup against Parliament in support of President MacMahon, a former Marshal. The military had its own fears, which were considerably more justified. Responding to constituent demands, republican politicians gradually worked to reduce the length of military service from seven to three years and to break down the quasi-monastic barriers insulating the regiment from secular, democratic trends in French society at large. Military professionals, while not averse to all reform, rightly feared a slippery slope towards a virtual militia system, in which the professional standing army would degenerate into a school for the superficial, short-term training of France's decidedly unmilitary youth. War college professors and military publicists like Georges Gilbert, responding to this danger, began by the 1880s to promote an offensive operational doctrine, which they claimed could only be implemented by well-trained, active-duty troops.[51]

This explosive situation was well managed by nationalist republican leaders like Léon Gambetta, leader of the French national resistance in the second phase of the Franco–Prussian War, and especially Charles de Freycinet, organizer of Gambetta's improvised popular armies. As War Minister in the 1880s and 1890s, Freycinet defused military fears and won their acceptance of the three-year service. He backed the military on questions of matériel, autonomy in matters of military justice, and selection of commanders on the basis of professional competence rather than political acceptability. At the same time, he pressed for more extensive use of the large pool of reservist manpower that was being created by the three-year conscription system, and the military was reasonably accommodating. In this context of moderate civil-military relations, war plans and doctrine were also moderate. Henri Bonnal's

50. Presenting somewhat contrasting views of French civil-military relations during this period are Douglas Porch, *March to the Marne: The French Army, 1871–1914* (Cambridge: Cambridge University Press, 1981) and David B. Ralston, *The Army of the Republic: The Place of the Military in the Political Evolution of France, 1871–1914* (Cambridge: M.I.T. Press, 1967).
51. See, for example, Gilbert, *Essais*, p. 271.

"defensive-offensive" school was the Establishment doctrine, reflected in the cautious, counteroffensive war plans of that era.[52]

Freycinet and other republican statesmen of the militant neo-Jacobin variety cherished the army as the instrument of revanche and as a truly popular institution, with roots in the *levée en masse* of the Wars of the Revolution. Though he wanted to democratize the army, Freycinet also cared about its fighting strength and morale, unlike many later politicians who were concerned only to ease their constituents' civic obligations. His own moderate policies, respectful of military sensitivities but insistent on key questions of civilian control, elicited a moderate response from military elites, whose propensity to develop a self-protective organizational ideology was thus held in check.

The deepening of the Dreyfus crisis in 1898 rekindled old fears on both sides and destroyed the system of mutual respect and reassurance constructed by Freycinet. The military's persistence in a blatant miscarriage of justice against a Jewish General Staff officer accused of espionage confirmed the republicans' view of the army as a state within the state, subject to no law but the reactionary principles of unthinking obedience and blind loyalty. When conservatives and monarchists rallied to the military's side, it made the officer corps appear (undeservedly) to be the spearhead of a movement to overthrow the Republic. Likewise, attacks by the Dreyfusards confirmed the worst fears of the military. Irresponsible Radicals were demanding to meddle in the army's internal affairs, impeaching the integrity of future wartime commanders, and undermining morale. Regardless of Dreyfus's guilt or innocence, the honor of the military had to be defended for the sake of national security.

The upshot of the affair was a leftward realignment of French politics. The new Radical government appointed as War Minister a young reformist general, Louis André, with instructions to "republicanize" the army. André, aided by an intelligence network of Masonic Lodges, politicized promotions and war college admissions, curtailed officers' perquisites and disciplinary powers, and forced Catholic officers to participate in inventorying church property. In 1905, the term of conscription was reduced to two years, with reservists intended to play a more prominent role in war plans, field exercises, and the daily life of the regiment.

52. Charles de Freycinet, *Souvenirs, 1878–1893* (New York: Da Capo, 1973).

In this hostile environment, a number of officers—especially the group of "Young Turks" around Colonel Loyzeaux de Grandmaison—began to reemphasize in extreme form the organizational ideology propounded earlier by Gilbert. Its elements read like a list of the errors of Plan 17: *offensive à outrance,* mystical belief in group *élan* achieved by long service together, denigration of reservists, and disdain for reactive war plans driven by intelligence estimates. Aided by the Agadir Crisis of 1911, General Joffre and other senior figures seeking a reassertion of professional military values used the Young Turks' doctrine to scuttle the reformist plans of the "republican" commander in chief, Victor Michel, and to hound him from office. Michel, correctly anticipating the Germans' use of reserve corps in the opening battles and the consequent extension of their right wing across northern Belgium, had sought to meet this threat by a cordon defense, making intensive use of French reservists. Even middle-of-the-road officers considered ruinous the organizational changes needed to implement this scheme. It was no coincidence that Grandmaison's operational doctrine provided a tool for attacking Michel's ideas point-by-point, without having to admit too blatantly that it was the institutional implications of Michel's reservist-based plan that were its most objectionable aspect.[53] Having served to oust Michel in 1911, the Grandmaison doctrine also played a role (along with the trumped-up scenario of a German standing-start attack) in justifying a return to the three-year term of service in 1913. The problem was that this ideology, so useful as a tool for institutional defense, became internalized by the French General Staff, who based Plan 17 on its profoundly erroneous tenets.

Obviously, there is much that is idiosyncratic in the story of the *offensive à outrance.* The overlapping of social and civil-military cleavages, which produced an unusually intense threat to the "organizational essence" and autonomy of the French army, may have no close analog in the contemporary era. At a higher level of abstraction, however, a broadly applicable hypothesis may nonetheless be gleaned from the French experience. That is, doctrinal bias is likely to become more extreme whenever strategic doctrine can be used an an ideological weapon to protect the military organization from threats to its institutional interests. Under such circumstances, doctrine be-

53. An internal General Staff document that was highly critical of Michel's scheme stated: "It is necessary only to remark that this mixed force would require very profound changes in our regulations, our habits, our tactical rules, and the organization of our staffs." Cited in Snyder, *Ideology of the Offensive,* chapter 3.

comes unhinged from strategic reality and responds primarily to the more pressing requirements of domestic and intragovernmental politics.

Russia: Institutional Pluralism and Strategic Overcommitment

Between 1910 and 1912, Russia changed from an extremely cautious defensive war plan to an overcommitted double offensive against both Germany and Austria. The general direction of this change can be easily explained in terms of rational strategic calculations. Russia's military power had increased relative to Germany's, making an offensive more feasible, and the tightening of alliances made it more obvious that Germany would deploy the bulk of its army against France in the first phase of the fighting, regardless of the political circumstances giving rise to the conflict. Russian war planners consequently had a strong incentive to invade Germany or Austria during the "window of opportunity" provided by the Schlieffen Plan. Attacking East Prussia would put pressure on Germany's rear, thus helping France to survive the onslaught; attacking the Austrian army in Galicia might decisively shift the balance of power by knocking Germany's ally out of the war, while eliminating opposition to Russian imperial aims in Turkey and the Balkans.[54]

What is harder to explain is the decision to invade both Germany and Austria, which ensured that neither effort would have sufficient forces to achieve its objectives. At a superficial level the explanation for this failure to set priorities is simple enough: General Yuri Danilov and the General Staff in St. Petersburg wanted to use the bulk of Russia's forces to attack Germany, while defending against Austria; General Mikhail Alekseev and other regional commanders wanted to attack Austria, leaving a weak defensive screen facing East Prussia. Each faction had powerful political connections and good arguments. No higher arbiter could or would choose between the contradictory schemes, so a *de facto* compromise allowed each to pursue its preferred offensive with insufficient forces. At this level, we have a familiar tale of bureaucratic politics producing an overcommitted, Christmas-tree "resultant."[55]

---

54. Apart from Zaionchkovskii, the most interesting work on Russian strategy is V.A. Emets, *Ocherki vneshnei politiki Rossii v period pervoi mirovoi voiny: vzaimootnosheniia Rossii s soiuznikami po voprosam vedeniia voiny* (Moscow: Nauka, 1977).
55. On the characteristics of compromised policy, see Warner Schilling, "The Politics of National Defense: Fiscal 1950," in Schilling et al., *Strategy, Politics, and Defense Budgets* (New York: Columbia University Press, 1962), pp. 217–218.

At a deeper level, however, several puzzles remain. One is that "where you sat" bureaucratically was only superficially related to "where you stood" on the question of strategy. Alekseev was the Chief-of-Staff-designate of the Austrian front, so had an interest in making his turf the scene of the main action. But Alekseev had always preferred an Austria-first strategy, even when he had been posted to the General Staff in St. Petersburg. Similarly, Danilov served under General Zhilinskii, the Chief of Staff who negotiated a tightening of military cooperation with France after 1911, so his bureaucratic perspective might explain his adoption of the Germany-first strategy that France preferred. But Danilov's plans had always given priority to the German front, even in 1908–1910 when he doubted the reliability and value of France as an ally.[56] Thus, this link between bureaucratic position and preferred strategy was mostly spurious.

Bureaucratic position does explain why Alekseev's plan attracted wide support among military district chiefs of staff, however. These regional planners viewed the coming war as a problem of battlefield operations, not grand strategy. Alekseev's scheme was popular with them, because it proposed clear lines of advance across open terrain. Danilov's plans, in contrast, were a source of frustration for the commanders who would have to implement them. His defensive 1910 plan perplexed them, because it offered no clear objectives.[57] His 1913 plan for an invasion of East Prussia entailed all sorts of operational difficulties that local commanders would have to overcome: inordinate time pressure, the division of the attacking force by the Masurian Lakes, and the defenders' one-sided advantages in rail lines, roads, fortifications, and river barriers.

Nonetheless, the main differences between Danilov and Alekseev were intellectual, not bureaucratic.[58] Danilov was fundamentally pessimistic about Russia's ability to compete with modern, efficient Germany. He considered Russia too weak to indulge in imperial dreams, whether against Austria or Turkey, arguing that national survival required an absolute priority be given to containing the German danger. In 1910, this pessimism was expressed in his ultra-defensive plan, based on the fear that Russia would have to face Germany virtually alone. By 1913–1914, Danilov's pessimism took a different form. The improved military balance, the tighter alliance with France after

---

56. Zaionchkovskii, *Podgotovka Rossii k imperialisticheskoi voine*, pp. 184–190.
57. Ibid., pp. 206–207.
58. See Schilling, "Politics of National Defense," for this distinction.

Agadir, and telling criticism from Alekseev convinced Danilov that a porcupine strategy was infeasible politically and undesirable strategically. Now his nightmare was that France would succumb in a few weeks, once again leaving backward Russia to face Germany virtually alone. To prevent this, he planned a hasty attack into East Prussia, designed to draw German forces away from the decisive battle in France.

Alekseev was more optimistic about Russian prospects, supporting imperial adventures in Asia and anticipating that a "sharp rap" would cause Austria to collapse. Opponents of Danilov's Germany-first strategy also tended to argue that a German victory against France would be Pyrrhic. Germany would emerge from the contest bloodied and lacking the strength or inclination for a second round against Russia. A Russo–German condominium would ensue, paving the way for Russian hegemony over the Turkish Straits and in the Balkans.[59]

Available evidence is insufficient to explain satisfactorily the sources of these differing views. Personality differences may explain Danilov's extreme pessimism and Alekseev's relative optimism, but this begs the question of why each man was able to gain support for his view. What evidence exists points to idiosyncratic explanations: Danilov's plan got support from Zhilinskii (it fit the agreements he made with Joffre), the commander-designate of the East Prussian front (it gave him more troops), and the General Staff apparatus (a military elite disdainful of and pessimistic about the rabble who would implement their plans). Alekseev won support from operational commanders and probably from Grand Duke Nikolai Nikolaevitch, the future commander-in-chief and a quintessential optimist about Russian capabilities and ambitions. The War Minister, the Czar, and the political parties seem to have played little role in strategic planning, leaving the intramilitary factions to logroll their own disputes.[60]

Perhaps the most important question is why the outcome of the logrolling was not to scale down the aims of both offensives to fit the diminished forces available to each. In particular, why did Danilov insist on an early-start, two-pincer advance into East Prussia, when the weakness of each pincer made them both vulnerable to piecemeal destruction? Why not wait a few days

---

59. *Documents diplomatiques français (1871–1914),* Series 2, Vol. XII, p. 695, and other sources cited in Snyder, *Ideology of the Offensive,* chapter 7.
60. Norman Stone, *The Eastern Front, 1914–1917* (New York: Scribner's, 1975), chapter 1, presents some speculations about factional alignments, but evidence is inconclusive in this area.

until each pincer could be reinforced by late-arriving units, or why not advance only on one side of the lakes? The answer seems to lie in Danilov's extreme fears about the viability of the French and his consequent conviction that Russian survival depended on early and substantial pressure on the German rear. This task was a necessity, given his outlook, something that had to be attempted whether available forces were adequate or not. Trapped by his pessimism about Russia's prospects in the long run, Danilov's only way out was through unwarranted optimism about operational prospects in the short run. Like most cornered decision-makers, Danilov saw the "necessary" as possible.

This is an important theme in the German case as well. Schlieffen and the younger Moltke demonstrated an ability to be ruthlessly realistic about the shortcomings of their operational plans, but realism was suppressed when it would call into question their fundamental beliefs and values. Schlieffen's qualms about his war plan's feasibility pervade early drafts, but disappear later on, without analytical justification. He entertained doubts as long as he thought they would lead to improvements, but once he saw that no further tinkering would resolve the plan's remaining contradictions, he swept them under the rug. The younger Moltke did the same thing, resorting to blithe optimism only on make-or-break issues, like the seizure of Liège, where a realistic assessment of the risks would have spotlighted the dubiousness of *any* strategy for rapid, decisive victory. Rather than totally rethink their strategic assumptions, which were all bound up with fundamental interests and even personal characteristics, all of these strategists chose to see the "necessary" as possible.[61]

Two hypotheses emerge from the Russian case. The first points to bureaucratic logrolling as a factor that is likely to exacerbate the normal offensive bias of military organizations. In the absence of a powerful central authority, two factions or suborganizations will each pursue its own preferred offensive despite a dramatic deficit of available forces. Thus, offensives that are moderately ambitious when considered separately become extremely overcommitted under the pressure of scarce resources and the need to logroll with

61. Groener, writing in the journal *Wissen und Wehr* in 1927, p. 532, admitted that it had been mere "luck" that an "extremely important" tunnel east of Liège was captured intact by the Germans in August 1914. Ritter, *Schlieffen Plan*, p. 166, documents Moltke's uncharacteristic optimism about quickly seizing Liège and avoiding the development of a monumental logistical bottleneck there. In the event, the Belgians actually ordered the destruction of their bridges and rail net, but the orders were not implemented systematically.

other factions competing for their allocation. The German case showed how the lack of civilian control can produce doctrinal extremism when the military is united; the Russian case shows how lack of civilian control can also lead to extreme offensives when the military is divided.

The second hypothesis, which is supported by the findings of cognitive theory, is that military decision-makers will tend to overestimate the feasibility of an operational plan if a realistic assessment would require forsaking fundamental beliefs or values.[62] Whenever offensive doctrines are inextricably tied to the autonomy, "essence," or basic worldview of the military, the cognitive need to see the offensive as possible will be strong.

External Influences on Strategy and Civil-Military Relations

The offensive strategies of 1914 were largely domestic in origin, rooted in bureaucratic, sociopolitical, and psychological causes. To some extent, however, external influences exacerbated—and occasionally diminished—these offensive biases. Although these external factors were usually secondary, they are particularly interesting for their lessons about sources of leverage over the destabilizing policies of one's opponents. The most important of these lessons—and the one stressed by Van Evera elsewhere in this issue—is that offense tends to promote offense and defense tends to promote defense in the international system.

One way that offense was exported from one state to another was by means of military writings. The French discovered Clausewitz in the 1880s, reading misinterpretations of him by contemporary German militarists who focused narrowly on his concept of the "decisive battle." At the same time, reading the retrograde Russian tactician Dragomirov reinforced their homegrown overemphasis on the connection between the offensive and morale. Russian writings later reimported these ideas under the label of *offensive à outrance,* while borrowing from Germany the short-war doctrine. Each of Europe's militaries cited the others in parroting the standard lessons drawn from the Russo–Japanese War: offense was becoming tactically more difficult but was still advantageous strategically. None of this shuffling and sharing of rationales for offense was the initial cause of anyone's offensive bias. Everyone was exporting offense to everyone else; no one was just receiving.

62. Irving Janis and Leon Mann, *Decision Making* (New York: Free Press, 1977).

Its main effect was mutual reinforcement. The military could believe (and argue to others) that offense must be advantageous, since everyone else said so, and that the prevalence of offensive doctrines was somebody else's fault.[63]

The main vehicle for exporting offensive strategies was through aggressive policies, not offensive ideas. The aggressive diplomacy and offensive war plans of one state frequently encouraged offensive strategies in neighboring states both directly, by changing their strategic situation, and indirectly, by changing their pattern of civil-military relations. German belligerence in the Agadir crisis of 1911 led French civilians to conclude that war was likely and that they had better start appeasing their own military by giving them leaders in which they would have confidence. This led directly to Michel's fall and the rise of Joffre, Castelnau, and the proponents of the *offensive à outrance.* German belligerence in the Bosnian crisis of 1908–1909 had a similar, if less direct effect on Russia. It convinced Alekseev that a limited war against Austria alone would be impossible, and it put everyone in a receptive mood when the French urged the tightening of the alliance in 1911.[64] Before Bosnia, people sometimes thought in terms of a strategic modus vivendi with Germany; afterwards, they thought in terms of a breathing spell while gaining strength for the final confrontation. Combined with the Russians' growing realization of the probable character of the German war plan, this led inexorably to the conclusions that war was coming, that it could not be limited, and that an unbridled offensive was required to exploit the window of opportunity provided by the Schlieffen Plan's westward orientation. Caught in this logic, Russian civilians who sought limited options in July 1914 were easily refuted by Danilov and the military. Completing the spiral, the huge Russian arms increases provoked by German belligerence allowed the younger Moltke to argue persuasively that Germany should seek a pretext for preventive war before those increases reached fruition in 1917. This recommendation was persuasive only in the context of the Schlieffen Plan, which made Germany look weaker than it really was by creating needless enemies and wasting troops on an impossible task. Without the Schlieffen Plan, Germany would not have been vulnerable in 1917.

In short, the European militaries cannot be blamed for the belligerent diplomacy that set the ball rolling towards World War I. Once the process began, however, their penchant for offense and their quickness to view war

---

63. Snyder, *Ideology of the Offensive,* chapters 2 and 3.
64. Ibid., chapter 7, citing Zaionchkovskii, pp. 103, 350, and other sources.

as inevitable created a slide towards war that the diplomats did not foresee.[65] The best place to intervene to stop the destabilizing spiral of exported offense was, of course, at the beginning. If German statesmen had had a theory of civil-military relations and of the security dilemma to help them calculate risks more accurately, their choice of a diplomatic strategy might have been different.

If offense gets exported when states adopt aggressive policies, it also gets exported when states try to defend themselves in ways that are indistinguishable from preparations for aggression.[66] In the 1880s, the Russians improved their railroads in Poland and increased the number of troops there in peacetime, primarily in order to decrease their vulnerability to German attack in the early weeks of a war. The German General Staff saw these measures as a sign that a Russian attack was imminent, so counseled launching a preventive strike before Russian preparations proceeded further. Bismarck thought otherwise, so the incident did not end in the same way as the superficially similar 1914 case. Several factors may account for the difference: Bismarck's greater power over the military, his lack of interest in expansion for its own sake, and the absence of political conditions that would make war seem inevitable to anyone but a General Staff officer. Perhaps the most important difference, however, was that in 1914 the younger Moltke was anticipating a future of extreme vulnerability, whereas in 1887 the elder Moltke was anticipating a future of strategic stalemate. Moltke, planning for a defense in the west in any event, believed that the Germans could in the worst case hold out for 30 years if France and Russia forced war upon them.[67]

Although states can provoke offensive responses by seeming too aggressive, they can also invite offensive predation by seeming too weak. German hopes for a rapid victory, whether expressed in the eastward plan of the 1880s or the westward Schlieffen Plan, always rested on the slowness of Russia's mobilization. Likewise, Germany's weakness on the eastern front, artificially created by the Schlieffen Plan, promoted the development of offensive plans in Russia. Finally, Belgian weakness allowed the Germans to

---

65. Isabel V. Hull, *The Entourage of Kaiser Wilhelm II, 1888–1918* (New York: Cambridge University Press, 1982), discusses the effect on the Kaiser of his military aides' incessant warnings that war was inevitable.
66. Robert Jervis, "Cooperation under the Security Dilemma," *World Politics,* Vol. 31, No. 2 (January 1978), pp. 199–210.
67. Barbara Tuchman, *The Guns of August* (1962; rpt., New York: Dell, 1971), p. 38; see also *Aufmarschpläne,* pp. 150–156, for Moltke's last war plan of February 1888.

retain their illusions about decisive victory by providing an apparent point of entry into the French keep.

States who want to export defense, then, should try to appear neither weak nor aggressive. The French achieved this in the early 1880s, when a force posture heavy on fortifications made them an unpromising target and an ineffective aggressor. In the short run, this only redirected Moltke's offensive toward a more vulnerable target, Russia. But by 1888–1890, when Russia too had strengthened its fortifications and its defensive posture in Poland generally, Moltke was stymied and became very pessimistic about offensive operations. Schlieffen, however, was harder to discourage. When attacking Russia became unpromising, he simply redirected his attention towards France, pursuing the least unpromising offensive option. For hard core cases like Schlieffen, one wonders whether any strategy of non-provocative defense, no matter how effective and non-threatening, could induce abandoning the offensive.

Soviet Strategy and Civil-Military Relations

In 1914, flawed civil-military relations exacerbated and liberated the military's endemic bias for offensive strategies, creating strategic instability despite military technologies that aided the defender of the status quo. Some of the factors that produced this outcome may have been peculiar to that historical epoch. The full professionalization of military staffs had been a relatively recent development, for example, and both civilians and military were still groping for a satisfactory *modus vivendi*. After the First World War, military purveyors of the "cult of the offensive" were fairly well chastened except in Japan, where the phenomenon was recapitulated. Our own era has seen nothing this extreme, but more moderate versions of the military's offensive bias are arguably still with us. It will be worthwhile, therefore, to reiterate the kinds of conditions that have intensified this bias in the past in order to assess the likelihood of their recurrence.

First, offensive bias is exacerbated when civilian control is weak. In Germany before 1914, a long period of military autonomy in strategic planning allowed the dogmatization of an offensive doctrine, rooted in the parochial interests and outlook of the General Staff. In Russia, the absence of firm, unified civilian control fostered logrolling between two military factions, compounding the offensive preferences exhibited by each. Second, offensive bias grows more extreme when operational doctrine is used as a weapon in civil-military disputes about domestic politics, institutional arrangements, or other nonstrategic issues. The French *offensive à outrance*, often dismissed as some mystical aberration, is best explained in these terms.

Once it appears, an acute offensive bias tends to be self-replicating and resistant to disconfirming evidence. Offensive doctrinal writings are readily transmitted across international boundaries. More important, offensive strategies tend to spread in a chain reaction, since one state's offensive tends to create impending dangers or fleeting opportunities for other states, who must adopt their own offensives to forestall or exploit them. Finally, hard operational evidence of the infeasibility of an offensive strategy will be rationalized away when the offensive is closely linked to the organization's "essence," autonomy, or fundamental ideology.

I believe that these findings, derived from the World War I cases, resonate strongly with the development of Soviet nuclear strategy and with certain patterns in the U.S.–Soviet strategic relationship. At a time when current events are stimulating considerable interest in the state of civil-military relations in the Soviet Union, the following thoughts are offered not as answers but as questions that researchers may find worth considering.

Soviet military doctrine, as depicted by conventional wisdom, embodies all of the desiderata typically expressed in professional military writings throughout the developed world since Napoleon. Like Schlieffen's doctrine, it stresses offense, the initiative, and decisive results through the annihilation of the opponent's ability to resist. It is suspicious of political limitations on violence based on mutual restraint, especially in nuclear matters. Both in style and substance, Sidorenko reads like a throwback to the military writers of the Second Reich, warning that "a forest which has not been completely cut down grows up again."[68] The similarity is not accidental. Not only does offense serve some of the same institutional functions for the Soviet military as it did for the German General Staff, but Soviet doctrine is to some degree their lineal descendant. "In our military schools," a 1937 Pravda editorial averred, "we study Clausewitz, Moltke, Schlieffen, and Ludendorff."[69] Soviet nuclear doctrine also parallels pre-1914 German strategy in that both cut against the grain of the prevailing technology. The Soviets have never been

68. Quoted by Benjamin Lambeth, "Selective Nuclear Options and Soviet Strategy," in Johan Holst and Uwe Nerlich, *Beyond Nuclear Deterrence* (New York: Crane, Russak, 1977), p. 92.
69. Raymond Garthoff, *Soviet Military Doctrine* (Glencoe, Ill.: Free Press, 1953), p. 56.

in a position to achieve anything but disaster by seizing the initiative and striving for decisive results; neither was Schlieffen.

There are also parallels in the political and historical circumstances that permitted the development of these doctrines. The Soviet victories in World War II, like the German victories in 1866 and 1870, were nation-building and regime-legitimating enterprises that lent prestige and authority to the military profession, notwithstanding Stalin's attempt to check it. This did not produce a man on horseback in either country, nor did it allow the military to usurp authority on questions of the use of force. But in both cases the military retained a monopoly of military operational expertise and was either never challenged or eventually prevailed in practical doctrinal disputes. In the German case, at least, it was military autonomy on questions of operational plans and doctrine that made war more likely; direct lobbying for preventive strikes caused less trouble because it was clearly illegitimate.

While many accounts of the origins of Soviet nuclear strategy acknowledge the effect of the professional military perspective, they often lay more stress on civilian sources of offensive, warfighting doctrines: for example, Marxism–Leninism, expansionist foreign policy goals, and historical experiences making Russia a "militarized society." Political leaders, in this view, promote or at least accept the military's warfighting doctrine because it serves their foreign policy goals and/or reflects a shared view of international politics as a zero-sum struggle. Thus, Lenin is quoted as favoring a preemptive first strike, Frunze as linking offense to the proletarian spirit. The military principle of annihilation of the opposing armed force is equated with the Leninist credo of *kto kogo*.[70]

Although this view may capture part of the truth, it fails to account for recurrent statements by Soviet political leaders implying that nuclear war is unwinnable, that meaningful damage limitation cannot be achieved through superior warfighting capabilities, and that open-ended expenditures on strategic programs are wasteful and perhaps pointless. These themes have been voiced in the context of budgetary disputes (not just for public relations purposes) by Malenkov, Khrushchev, Brezhnev, and Ustinov. To varying degrees, all of these civilian leaders have chafed at the cost of open-ended warfighting programs and against the redundant offensive capabilities de-

70. Herbert Dinerstein, *War and the Soviet Union* (New York: Praeger, 1962), pp. 210–211; Garthoff, *Soviet Military Doctrine*, pp. 65, 149.

manded by each of several military suborganizations. McNamara discovered in the United States that the doctrine of mutual assured destruction, with its emphasis on the irrelevance of marginal advantages and the infeasibility of counterforce damage-limitation strategies, had great utility in budgetary debates. Likewise, recent discussions in the Soviet Union on the feasibility of victory seem to be connected with the question of how much is enough. Setting aside certain problems of nuance and interpretation, a case can be made that the civilian leadership, speaking through Defense Minister Ustinov, has been using strategic doctrine to justify slowing down the growth of military spending. In the context of arguments about whether the Reagan strategic buildup will really make the Soviet Union more vulnerable, Ustinov has quite clearly laid out the argument that neither superpower can expect to gain anything by striking first, since both have survivable retaliatory forces and launch-on-warning capabilities. Thus, Ustinov has been stressing that the importance of surprise is diminishing and that "preemptive nuclear strikes are alien to Soviet military doctrine." Ogarkov, the Chief of the General Staff, has been arguing the opposite on all counts: the U.S. buildup is truly threatening, the international scene is akin to the 1930s, the surprise factor is growing in importance, damage limitation is possible (though "victory" is problematic), and consequently the Soviet Union must spare no expense in preparing to defend itself.[71]

This is somewhat reminiscent of the French case in World War I, in which civilians and the military were using doctrinal arguments as weapons in disputes on other issues. Two related dangers arise in such situations. The first is that doctrinal argumentation and belief, responding to political and organizational necessity, lose their anchoring in strategic realities and become dogmatic and extremist. The second is that a spiral dynamic in the political dispute may carry doctrine along with it. That is, the harder each side fights to prevail on budgetary or organizational questions, the more absolute and unyielding their doctrinal justifications will become. In this regard, it would be interesting to see whether the periods in which Soviet military spokesmen

71. Citations to the main statements by Ogarkov and Ustinov can be found in Dan L. Strode and Rebecca V. Strode, "Diplomacy and Defense in Soviet National Security Policy," *International Security,* Vol. 8, No. 2 (Fall 1983), pp. 91–116. Quotation from William Garner, *Soviet Threat Perceptions of NATO's Eurostrategic Missiles* (Paris: Atlantic Institute for International Affairs, 1983), p. 69, citing *Pravda,* July 25, 1981. I have benefitted from discussions of the Ogarkov and Ustinov statements with Lawrence Caldwell, Stephen Coffey, Clifford Kupchan, and Cynthia Roberts, who advanced a variety of interpretations not necessarily similar to my own.

were arguing hardest that "victory is possible" coincided with periods of sharp budgetary disputes.

Even if some of the above is true, the pattern may be a weak one in comparison with the French case. Ustinov is more like Freycinet than André, and marginal budgetary issues do not carry the same emotional freight as the threats to organizational "essence" mounted in the Dreyfus aftermath. Still, if we consider that the Soviet case couples some of the autonomy problems of the German case with some of the motivational problems of the French case, a volatile mixture may be developing.

Another civil-military question is whether Soviet military doctrine is mismatched with Soviet diplomacy. On the surface, it may seem that the awe-inspiring Soviet military machine and its intimidating offensive doctrine are apt instruments for supporting a policy of diplomatic extortion. It may, however, pose the same problem for Soviet statesmen that the Schlieffen Plan did for Bülow and Bethmann. Soviet leaders may be self-deterred by the all-or-nothing character of their military options.[72] Alternatively, if the Soviets try to press ahead with a diplomacy based on the "Bolshevik operational code" principles of controlled pressure, limited probes, and controlled, calculated risks, they may find themselves trapped by military options that create risks which cannot be controlled.

These problems may not arise, however, since the Soviets seem to have turned away from Khrushchev's brinkmanship diplomacy. In the Brezhnev era, Soviet doctrine on the political utility of nuclear forces stressed its role as an umbrella deterring intervention against "progressive" political change.[73] Insofar as limited options and "salami tactics" are more clearly indispensable for compellent than for deterrent strategies, this would help to solve the Soviet diplomats' mismatch problem. The "last clear chance" to avoid disaster would be shifted onto the United States. This solution to the diplomats'

---

72. Increased Soviet attention to the "conventional option" since the late 1960s would seem to have mitigated this problem, but in fact it may have compounded it. Military interest in preparing for a conventional phase and acquiring capabilities for escalation dominance in the theater may derive more from obvious organizational motives than from a fundamental change in the military's mind-set of "inflexible over-response." In Soviet thinking, limitations seem to be based less on mutual restraint than on NATO's willingness to see its theater nuclear forces destroyed during the conventional phase. This raises the nightmarish possibility that the Soviet leadership could embark on war thinking that it had a conventional option, whereas in fact unrestrained conventional operations and preemptive incentives at the theater nuclear level would lead to rapid escalation.

73. Coit Blacker, "The Kremlin and Detente: Soviet Conceptions, Hopes, and Expectations," in Alexander George, ed., *Managing U.S.–Soviet Rivalry* (Boulder: Westview, 1983), pp. 122–123.

problem might cause problems for the military's budget rationale, however, since strategic parity should be sufficient to carry out a strictly deterrent function.

The German case suggests that extremism in strategic thinking may depend a great deal on institutionalization and dogmatization of doctrine in the mature military organization. If Roman Kolkowicz's "traditionalists" are equated with the Moltke generation and his "modernist" technocrats with the Schlieffen generation, do we find a parallel in the dogmatization of doctrine? Benjamin Lambeth argues that Soviet doctrine is quite flexible and creative, but so was Schlieffen on questions of how to implement his strategic tenets under changing conditions.[74] Creativity within the paradigm of decisive, offensive operations may coexist with utter rigidity towards options that would require a change in the basic paradigm. For example, the Soviet ground forces adapted creatively to improvements in precision-guided munitions (PGMs) that seemed to threaten the viability of their offensive doctrine; they did not consider, however, that PGMs might offer an opportunity to give up their fundamentally offensive orientation. As for the third phase of organizational evolution, are there any parallels to Ludendorff or Groener among younger Soviet officers? Are they forging links to Russian nationalists, whose social base Alexander Yanov describes in ways that are strongly reminiscent of Eley's account of the ultranationalist German right?[75]

Any discussion of the extremist potential of Soviet strategy must consider the strong reality constraint imposed by the mutual-assured-destruction relationship. Despite the reckless rhetoric of some junior officers, it seems clear that when the head of the Strategic Rocket Forces said in 1967 that "a sudden preemptive strike cannot give [the aggressor] a decisive advantage," he knew that launch-on-warning and the hardening of silos made this true for both sides.[76] And today Ogarkov does not deny that a scot-free victory is impossible. But despite this, the theme of damage limitation remains strong in Soviet military thinking, and we should remember those World War I strategists who saw the "necessary" as possible, no matter how realistically they did their operational calculations.

74. Lambeth, "Selective Nuclear Options"; Kolkowicz, *The Soviet Military and the Communist Party* (Princeton: Princeton University Press, 1967).
75. Alexander Yanov, *Detente after Brezhnev* (Berkeley: Institute of International Studies, University of California, 1977).
76. Garner, *Soviet Threat Perceptions*, p. 69.

Finally, how have the policies of the United States affected the development of civil-military relations and strategic doctrine in the U.S.S.R.? Some analysts argue that the Ogarkov–Ustinov debates ended in May 1983 with Ustinov's capitulation, at least on the level of rhetoric. Although leadership politics may have been a factor, a more important reason may have been the Reagan "Star Wars" speech and the Reagan defense program generally.[77] Echoing the developments in France in 1911, rising levels of external threat may have helped the military to win the doctrinal argument and achieve its institutional aims in the underlying issues tied to the doctrinal dispute. This episode may also be seen as the latest round of a process of exporting and re-importing warfighting strategies. The impact of Soviet counterforce doctrines on the American strategic debate in the 1970s is obvious; now the fruits of our conversion are perhaps being harvested by Ogarkov in Soviet debates on military budgets and operational policies.

Whatever the precise reality of current civil-military relations in the Soviet Union, patterns revealed by the World War I cases suggest that the Soviet Union manifests several "risk factors" that could produce an extreme variant of the military's endemic offensive bias. The historical parallel further suggests that the actions of rival states can play an important role in determining how these latent risks unfold. Aggressive policies were liable to touch off these latent dangers, but vulnerability also tended to encourage the opponent to adopt an offensive strategy. Postures that were both invulnerable and non-provocative got the best results, but even these did not always dissuade dogmatic adherents to the "cult of the offensive." Although Soviet persistence in working the problems of conventional and nuclear offensives does recall the dogged single-mindedness of a Schlieffen, nuclear weapons pose a powerful reality constraint for which no true counterpart existed in 1914. Consequently, if the twin dangers of provocation and vulnerability are avoided, there should be every hope of keeping Soviet "risk factors" under control. The current drift of the strategic competition, however, makes that not a small "if."

---

77. Setting these debates into the context of U.S.–Soviet relations are Lawrence T. Caldwell and Robert Legvold, "Reagan Through Soviet Eyes," *Foreign Policy*, No. 52 (Fall 1983), pp. 3–21.

Windows of Opportunity

Do States Jump Through Them?

Richard Ned Lebow

A "window of opportunity," a period during which a state possesses a significant military advantage over an adversary, has been a central concern of American strategic analysis. Central Intelligence Agency analysts regularly prepare National Intelligence Estimates (NIEs) of present and future Soviet strategic capabilities, with the primary objective of sounding the tocsin should a "window of vulnerability"—a window of opportunity from the perspective of the disadvantaged side—appear likely to open.[1] Several times in the past, American intelligence and defense experts have worried that such windows were about to open because of some reputed Soviet military capability; the "bomber gap," "missile gap," and the "ABM gap" were all predictions of this kind that failed to materialize.[2]

The author wishes to acknowledge the helpful comments and assistance of Matthew Evangelista, Alexander George, Irving Janis, Gert Krell, Herbert Lin, Jack Snyder, Janice Gross Stein, Douglas T. Stuart, and Stephen Van Evera.

Richard Ned Lebow is Professor of Government and Director of the Peace Studies Program at Cornell University.

---

1. For details about National Intelligence Estimates of Soviet Strategic Capabilities, see Lawrence Freedman, *U.S. Intelligence and the Soviet Strategic Threat* (London: Macmillan, 1977); John Prados, *The Soviet Estimate: U.S. Intelligence Analysis and Russian Military Strength* (New York: Dial Press, 1982).
2. Examples include: NSC-68, "United States Objectives and Programs for National Security," April 14, 1950, *Foreign Relations of the United States, 1950* (Washington, D.C.: U.S. Government Printing Office, 1977), Vol. 1, pp. 235–292, which spoke of the possibility of a Soviet attack by 1955 unless the United States initiated crash measures to build up its own strategic strength. The Killian Report, presented to the National Security Council in February 1955, warned that the Soviet Union might be tempted to initiate an atomic strike because of inadequate American early warning and defensive capabilities. It also warned of the possibility of a window of vulnerability after 1958 if the United States did not make as rapid progress as the Soviet Union in developing ICBMs. James R. Killian, Jr., *Sputniks, Scientists and Eisenhower* (Cambridge: M.I.T. Press, 1977), pp. 71–79. The Gaither Report, submitted to the National Security Council in 1957, emphasized even more than had the Killian Report, the danger of a window of vulnerability due to a possible missile gap. Security Resources Panel of the Scientific Advisory Committee (Gaither Committee), *Deterrence and Survival in the Nuclear Age* (Washington, D.C., November 1957). Declassified January 1973. For a historical treatment of American postwar window of vulnerability predictions, see Robert H. Johnson, "Periods of Peril: The Window of Vulnerability and Other Myths," *Foreign Affairs*, Vol. 61, No. 4 (Spring 1983), pp. 950–970.

International Security, Summer 1984 (Vol. 9, No. 1) 0162-2889/84/010147-40 $02.50/1

Since 1972, some prominent individuals within the American strategic community have been insisting that the Soviet Union would possess a window of opportunity in the early 1980s because of its significant advantage in powerful and accurate land-based missiles. Paul Nitze, William van Cleave, and others have worried aloud that Moscow would be tempted to launch a surprise attack during this period or, failing that, to exploit its advantage by pursuing a more "adventurist" foreign policy.[3] These strategic Cassandras urged a major expansion in American strategic forces, improvements in their command and control, and a series of short-term "fixes" to buttress American retaliatory capability during the early and middle 1980s. Even with such efforts, two well-known hawkish analysts asserted back in 1978 that "a healthy amount of luck will be required to emerge unscathed from what will be an irretrievably vulnerable period during the first half of the 1980's."[4] President Reagan has picked up this theme and made it a central justification of his call for new and more strategic weapons.[5]

The notion that the Soviets have a strategic advantage in the middle 1980s has been challenged principally on technical grounds.[6] Telling political criti-

---

3. For a sampling of this literature, see: Senate Armed Services Committee, *Hearings on Military Implications of the Treaty on the Limitation of Anti-Ballistic Missile Systems and the Interim Agreement on Limitation of Strategic Offensive Arms* (Washington, D.C.: U.S. Government Printing Office, 1972), pp. 145, 579, for the testimony of Senator Henry Jackson and William Van Cleave; Paul Nitze, "Deterring Our Deterrent," *Foreign Policy*, No. 25 (Winter 1976–77), pp. 195–210; Richard Pipes, "Why the Soviet Union Thinks It Could Fight and Win a Nuclear War," *Commentary*, July 1977, pp. 21–34; Colin S. Gray, "The Strategic Forces Triad: End of the Road," *Foreign Affairs*, Vol. 56, No. 5 (July 1978), pp. 771–789; William R. Graham and Paul Nitze, "Viable U.S. Strategic Missile Forces for the Early 1980's," in William R. Van Cleave and W. Scott Thompson, *Strategic Options for the Early Eighties* (New York: National Strategy Information Center, 1979), pp. 125–140; John J. Dziak, *Soviet Perceptions of Military Power* (New York: Crane, Russak, 1981), p. 20; Keith Payne, *Nuclear Deterrence in U.S.–Soviet Relations* (Boulder, Colo.: Westview, 1982), pp. 1–3; Myron Rush, "Guns over Growth in Soviet Policy," *International Security*, Vol. 7, No. 3 (Winter 1982–83), pp. 167–179.
4. T.K. Jones and W. Scott Thompson, "Central War and Civil Defense," *Orbis*, Vol. 22, No. 3 (Fall 1978), pp. 681–712. This thesis was even the premise of a best-selling spy novel: Douglas Terman, *First Strike* (New York: Pocket Books, 1978).
5. In a press conference in October 1981, Reagan warned that "a window of vulnerability is opening" that will jeopardize our hopes for "peace and freedom." Cited in Robert Scheer, *With Enough Shovels: Reagan, Bush and Nuclear War* (New York: Random House, 1982), p. 67. His most recent reference to a window of vulnerability was in his press conference on February 22, 1984 at which he attempted to justify his proposed defense budget as necessary to close that window. *The Washington Post*, February 23, 1984, p. 1.
6. Critics dispute the claim that the American land-based missile force is as vulnerable as proponents of the window of vulnerability thesis contend. They also question the operational reliability and accuracy of the Soviet missile force and emphasize the difficulty of coordinating a first strike and of overcoming the problem of "fratricide" associated with it. Skeptics further contend that destruction of the entire American land-based missile force in their silos, assuming

cisms also can be leveled at the window of vulnerability thesis. It is based on the premise that relative military advantage is the decisive consideration in a state's decision to go to war. Analysts who believe this are guilty of conflating means and ends. History indicates that wars rarely start because one side believes it has a military advantage. Rather, they occur when leaders become convinced that force is necessary to achieve important goals. War as Clausewitz observed—and the Soviets proclaim—is an extension of politics by other means. Its scope, strategy, and timing are determined, if often imperfectly, by the political objective for which the war is fought. Relative military advantage is merely one component of any decision regarding war. It is by no means even the most important one as examples can be drawn from every era of states' knowingly starting wars without a military advantage.[7]

Window of vulnerability arguments also tend to ignore the host of non-military factors which can influence leaders in the direction of peace. This article explores the importance of two such constraints: the absolute cost of war in human and economic terms and the personal reluctance of leaders to

---

for the moment that this could be done, would constitute the most pyrrhic of victories given the magnitude of the retaliatory capability that could be expected to survive at sea. Proponents of the window of vulnerability thesis counter with the assertion that the President would be deterred from retaliating with submarine-based missiles as they could only be used against Soviet cities and would invite the destruction of American cities in return. The critics scoff at this notion of "self-deterrence," arguing that a Soviet strike against U.S. missile and bomber bases and their command and control centers would kill so many Americans that the President would have little incentive (or ability) to withhold retaliation. To this, the window theorists reply that a Soviet counterforce strike could be carried out with relatively little collateral damage, killing perhaps fewer than five million Americans. Such claims nevertheless appear unrealistic in light of the findings of most studies of the effects of nuclear war. For some of the better technical critiques, see: Stanley Sienkiewicz, "Observations on the Impact of Uncertainty in Strategic Analysis," *World Politics*, Vol. 32 (October 1979), pp. 90–110; John D. Steinbruner and Thomas M. Garwin, "Strategic Vulnerability: The Balance Between Prudence and Paranoia," *International Security*, Vol. 1, No. 1 (Summer 1976), pp. 138–181; Alexander and Andrew Cockburn, "The Myth of Missile Accuracy," *New York Review of Books*, November 1980, pp. 40–44; J. Edward Anderson, "First Strike: Myth or Reality?," *Bulletin of the Atomic Scientists*, Vol. 37 (November 1981), pp. 6–11; Steve Smith, "MX and the Vulnerability of American Missiles," *Armament and Disarmament Information Unit Report*, Vol. 4 (May–June 1982), pp. 1–5; J. Edward Anderson, "Strategic Missiles Debated: Missile Vulnerability—What You Can't Know!," *Strategic Review*, Vol. 10 (Spring 1982), pp. 38–43 and the rebuttal by Robert T. Marsh, "Strategic Missiles Debated: Missile Accuracy—We Do Know!," pp. 35–37; Matthew Bunn and Kosta Tsipis, "The Uncertainties of a Preemptive Nuclear Attack," *Scientific American*, Vol. 249 (November 1983), pp. 38–47.

7. This argument is developed more fully by Richard Ned Lebow, "Misconceptions in American Strategic Assessment," *Political Science Quarterly*, Vol. 97 (Summer 1982), pp. 187–206, and Richard Ned Lebow, "Conclusions," to Robert Jervis, Richard Ned Lebow, and Janice Gross Stein, *Psychology and Deterrence* (Baltimore: Johns Hopkins University Press, forthcoming).

assume responsibility for it. It will attempt to document the ways in which these can discourage leaders from a war policy.

Three Windows of Opportunity

One way of analyzing the importance of nonmilitary considerations in decisions about peace and war would be to examine several instances in which peace was preserved even though a calculus of cost and gain of strategic interests ought to have pointed towards war. The most dramatic situations of this kind would be those in which policymakers decided against war with an apparently irreconcilable adversary despite their expectation that the military balance, seen at the time as highly favorable to themselves, would worsen dramatically in the near future. In keeping the peace they refused to exploit the advantages of a window of opportunity. Three twentieth-century examples come to mind: Germany in several crises prior to 1914, the United States in the 1950s, and the Soviet Union in the late 1960s.

Between 1891, the year of the Franco–Russian alliance, and the outbreak of war in 1914, Germany had a pronounced but gradually declining military edge over its two continental adversaries. The General Staff's solution to the problem posed by the prospect of a two-front war was the ill-fated Schlieffen Plan which aimed to defeat Germany's adversaries sequentially. The Plan was made possible by the slow rate of Russian mobilization which was expected to give Germany just enough time to overwhelm France by means of a lightning offensive before having to confront massive Russian armies on the eastern borders.[8]

From 1905 on, the German General Staff were increasingly concerned about the pace of Russian railway construction and military reforms. They feared that these activities would speed up Russian mobilization and make Russia's army more effective, thereby rendering the Schlieffen Plan unworkable. The generals repeatedly urged the Kaiser and the Chancellor to wage a preventive war against Russia and France while victory was still possible. In 1905, in 1909, and again in 1912, Germany's leaders rejected the pleas of the generals despite their recognition of the worsening military situation. In 1914, when Germany did go to war, it was the result of a crisis initiated by Austria; at

8. On the Schlieffen Plan, see Gerhard Ritter, *The Schlieffen Plan*, trans. Andrew and Eva Wilson (New York: Praeger, 1958); Lancelot L. Farrar, Jr., *The Short War: German Policy, Strategy, and Domestic Affairs, August–December, 1914* (Santa Barbara, Calif.: ABC-CLIO, 1973), pp. 10–33.

the outset neither the Kaiser nor the Chancellor expected that German support for Austria would lead to anything more than a localized Balkan conflict. What is surprising given the assumptions of deterrence theory and the window of vulnerability thesis is not that Germany stumbled into a European war in 1914 but that it consciously rejected such a war on three previous and more favorable occasions.

American restraint towards the Soviet Union during the years when Washington had a near monopoly in deliverable nuclear weapons constitutes a more recent example of a state's failure to exploit an apparent window of opportunity. Despite widespread recognition that the Soviets would sooner or later develop the means to make the continental United States vulnerable to attack, American leaders rejected the possibility of preventive war. They did this even though many officials feared that the Soviets would be tempted to start a nuclear war as soon as they acquired the strategic wherewithal to do so. Throughout this period, from roughly 1950 to 1965, American policymakers also refrained from attempting to exploit their nuclear superiority to roll back Soviet influences in central Europe, a policy rejected in practice because of the attendant risk of war. The Cold War rhetoric of Dulles and Kennedy aside, caution was the order of the day. Only in Cuba in 1962 did an American President deliberately pursue a policy that was seen at the time to court some risk of nuclear war. But this was done in circumstances in which, from Washington's perspective, the Americans were on the defensive and responding to a Soviet initiative which they believed to threaten the political and strategic status quo.

The Soviet Union evidenced similar self-restraint *vis à vis* its Chinese adversary in the late 1960s and the early 1970s. The Soviet image of China at that time as an implacable and irrational foe apparently led to serious consideration in Moscow in 1969–1970 of a preventive strike against Chinese nuclear facilities in order to deny Peking a nuclear capability. Soviet leaders decided against such a strike even though it meant accepting their own certain vulnerability in the not so distant future.[9]

While these three examples share obvious similarities, it is also important to recognize the differences among them. The Germans had to decide whether or not to start a conventional war, whereas the Americans and

9. Marvin Kalb and Bernard Kalb, *Kissinger* (Boston: Little, Brown, 1974), pp. 258–261; John Newhouse, *Cold Dawn: The Story of SALT* (New York: Holt, Rinehart and Winston, 1973), pp. 164, 188–189: Henry Kissinger, *White House Years* (Boston: Little, Brown, 1979), p. 183.

Russians confronted the more horrifying prospect of nuclear war. This, as we shall see, was a conditioning but not determining factor in their choice of policies. The Germans and the Americans would have fought an all-out war against their respective adversaries. The Russians, by contrast, presumably contemplated only a limited war against China. Most importantly of all, the Germans, alone among the three, actually went to war before their perceived window of opportunity closed. However, the extent to which 1914 represented a purposeful effort to exploit this window of opportunity is debatable.

All three examples probably fall short of fulfilling completely the parameters ascribed earlier to a classic window of opportunity situation. In the Russian case, we simply lack the information necessary to make an informed judgment. The German situation probably offers the best fit, but it is complicated by the fact that Berlin perceived its security dilemma more acutely in 1914 than it had in earlier years. Expansion of the Russian army and its stocks in 1913–1914, the tightening of the Entente, and the deepening of Austria's dilemma in the Balkans made the future look even dimmer to German leaders. For them, it could be argued, it was no longer a question of exploiting a window of opportunity but rather of forestalling a window of vulnerability. Strictly speaking, therefore, 1905 and 1912 were not fully comparable to 1914.[10]

In the American case, the principal divergence is the discrepancy between the years of the most fatalistic American expectations of the likelihood of war with the Soviet Union and the years of greatest American military advantage. By most accounts, American perceptions of the gravity of the international situation and, with it, expectations of future war reached two peaks: in 1948, in response to the Czech coup and the Berlin blockade, and in the summer of 1950 after the invasion of South Korea.[11] In the interim, the Russians

---

10. This question is treated in more detail by Stephen Van Evera and Jack Snyder in their contributions to this issue of *International Security*.

11. Harry Truman, who saw the Czech coup as a replay of Germany's aggression in the 1930s, wrote to his family on March 3, 1948: "We are faced with exactly the same situation with which Britain and France were faced in 1938–39 with Hitler. Things look black." Joint Chiefs of Staff records, cited in Gregg Herken, *The Winning Weapon: The Atomic Bomb in the Cold War, 1945–1950* (New Haven: Yale University Press, 1980), p. 246. Documents indicative of an equally fatalistic official American mood in the fall of 1948 are Foy D. Kohler to the secretary of state, September 28, 1948; Walter Bedell Smith to the secretary of state, December 23, 1948. *Foreign Relations of the United States, 1948*, Vol. 4: *Eastern Europe; The Soviet Union*. (Washington, D.C.: U.S. Government Printing Office, 1974), pp. 920, 946–947. George F. Kennan, in his *Memoirs*,

exploded their first atomic device, something American political leaders had not expected for years to come.[12] News of this event dramatically intensified American perceptions of the Soviet threat and for the first time imparted a real sense of urgency to it; policymakers were forced to recognize that it was only a matter of time before the Soviet Union produced enough bombs and long range bombers to put American cities at risk.[13] However, as the 1950s progressed, the sense of acute foreboding brought about by these developments gradually abated, defused initially by the death of Stalin and the end of the Korean War.

Although the United States possessed a nuclear monopoly from 1945 to 1949 and a near monopoly on intercontinentally deliverable weapons well into the late 1950s, the American nuclear advantage did not peak until about 1960. The reasons for this were a shortage of weapons and of the means to

---

1950–1963 (Boston: Little, Brown, 1972), pp. 91–92, remembers that in 1949–1950 most of Washington was convinced that the Russians would start a war as soon as their military preparations were completed—sometime perhaps in 1952. "They could not free themselves from the image of Hitler and his timetables. They viewed the Soviet leaders as absorbed with the pursuit of something called a 'grand design'—a design for the early destruction of American power and world conquest." A good discussion of American thinking about the prospect of war during the years 1948–1950 is provided by John Lewis Gaddis, "Was the Truman Doctrine a Real Turning Point?," *Foreign Affairs*, Vol. 52, No. 2 (January 1979), pp. 386–402, and *Strategies of Containment: A Critical Appraisal of Postwar American National Security Policy* (New York: Oxford University Press, 1982), pp. 89–126.

12. In 1945, American nuclear scientists had predicted that the Soviet Union could develop its own bomb within four to five years. The politicians, however, preferred not to give much credence to this professional consensus; it was apparently too threatening to them. The discovery, from rainwater samples collected in September 1949, that the Russians had exploded a bomb the month before, sent shudders through Washington. According to Herken, *Winning Weapon*, pp. 302–303, the initial reaction was one of "shocked disbelief." This was followed by attempts to deny the event, a sure sign of just how disturbing the reality of Soviet nuclear capability must have been. General Leslie Groves argued that the radiation must have been from a nuclear accident, not a test of a device. Lewis Strauss, *Men and Decisions* (Garden City, N.Y.: Doubleday, 1962), p. 206, reports that one cabinet officer reacted to the news of the test with the suggestion that the monitoring program be disbanded! J. Robert Oppenheimer recalls that Harry Truman refused to believe as late as 1953 that the Russians had either an atomic or hydrogen bomb. *The Open Mind* (New York: Simon and Schuster, 1955), p. 70.

13. "A-day," the time when the Russians would be ready to launch a nuclear attack against the United States, was thought likely to arrive sometime in 1952–1953. The Finletter Commission [*Survival in the Air Age: A Report by the President's Air Policy Commission* (Washington, D.C.: U.S. Government Printing Office, 1948)] set January 1, 1953 as the earliest possible date. Senator Stuart Symington told the Mahon Committee in 1950 that most civilians in the Defense Department thought that Moscow would not have sufficient weapons to conduct an attack before 1952. Cited in Warner R. Schilling, "The Politics of National Defense: Fiscal 1950," in Warner R. Schilling, Paul Y. Hammond, and Glenn H. Snyder, *Strategy, Politics and Defense Budgets* (New York: Columbia University Press, 1962), p. 83. George Kennan, *Memoirs*, Vol. 2, p. 91, reports that military planners, and to some extent political planners, had, against his objections, set 1952 as the probable peak danger period for war.

deliver them. In February 1948, there were only thirty-three B-29s in the entire American air force which were capable of carrying atomic bombs. By the end of the year the situation had improved somewhat. Operation Fleetwood, which became the official war plan in the fall of 1948, envisaged an atomic blitz using every weapon available at that time, 133. The attack was expected to kill almost seven million Soviet citizens and to destroy 40 percent of Soviet industry. The Joint Chiefs of Staff hoped that the "psychological shock" of such an assault might obviate the need for protracted warfare in Europe. Dissenters argued that Fleetwood would do little to blunt the Soviet capability to conduct offensive operations in Western Europe. American war planners did not feel secure about their ability to destroy the Soviet Union as a functioning state until well into the middle 1950s. By that time, they had hydrogen as well as atomic bombs in their arsenal and a growing fleet of B-47s and B-52s to deliver them. But as late as 1956, President Eisenhower spoke of a "decisive" nuclear strike against the Soviet Union as an option that the United States still did not possess.[14] It seems, therefore, that the United States did not really develop the means to conduct a successful preventive war until sometime after expectations of the imminent likelihood of war with the Soviet Union had significantly abated.

Despite the imperfect correspondence between these cases and what might be considered an "ideal type" window of opportunity situation, there is still enough of a fit to talk about the cases as constituting windows. In all three instances, the extreme hostility and aggressive character of the adversary was taken as a given. In all three, the state in question possessed a significant if not always overwhelming military advantage. And each of the adversaries was expected to improve his military capability in the near future to the point where that advantage was no longer decisive or even disappeared

14. On the development of nuclear capabilities and war plans in the 1940s and 1950s, see: Kenneth W. Condit, *The History of the Joint Chiefs of Staff*, Volume II: *1947–1949* (Wilmington, Delaware: Michael Glazier, 1979), pp. 283–309; David Alan Rosenberg, "The Origins of Overkill: Nuclear Weapons and American Strategy, 1945–1960," *International Security*, Vol. 7, No. 4 (Spring 1983), pp. 3–71, and specifically pp. 38–40, for details about Operation Fleetwood; George H. Quester, *Nuclear Diplomacy: The First Twenty-Five Years* (New York: Dunellen, 1970), pp. 1–8, for a discussion of the technical problems associated with preventive war in the late 1940s; Desmond Ball, *Targeting for Strategic Deterrence*, Adelphi Paper No. 185 (London: International Institute for Strategic Studies, 1983), pp. 3–17 for targeting plans during this period; David Alan Rosenberg, "U.S. Nuclear Stockpile, 1945 to 1950," *The Bulletin of the Atomic Scientists*, Vol. 38 (May 1982), pp. 25–30; Thomas B. Cochran, William M. Arkin, and Milton M. Hoenig, *Nuclear Weapons Databook*, Volume I: *U.S. Nuclear Forces and Capabilities* (Cambridge: Ballinger, 1984), pp. 6–12 for data on the historical growth of the United States nuclear arsenal.

altogether. Without undue exaggeration, proponents of preventive war could and did argue that their adversary was on the verge of acquiring the means of threatening the very survival of their state. Preventive war promised to ward off this military threat. It also held out the prospect of victory at less cost than a future war whose outcome would be uncertain. A decision based solely upon calculations of strategic cost and gain might well have dictated a decision in favor of war. This failed to happen, we must conclude, because of the salience of other, nonmilitary considerations in the minds of the policymakers in all three cases.

Absolute versus Relative Costs

The most obvious nonmilitary considerations are those of a political nature. Leaders who believe that their national interests require military action can still be constrained by public opinion or other domestic political forces. Prior to 1914, Austrian leaders were prevented by Hungarian opposition from starting a war with Serbia; only Germany's "blank check" and subsequent pleas for decisive action enabled them to overcome this obstacle in July 1914. Pacific or divided public opinion forestalled American intervention in the two world wars until either enough American ships had been sunk or the country itself had been attacked. French and British leaders were similarly constrained from acting against Hitler in the mid-1930s at a time when Germany might easily have been crushed. Ronald Reagan currently confronts public opposition to his administration's efforts to expand America's military role in Central America.

A second consideration militating against war can be its expected costs. In some circumstances they may be high enough to dissuade policymakers from using force regardless of the magnitude of the expected gains. Window of vulnerability arguments and deterrence theory in general conceive of political and military costs in relative terms. They describe the calculus of decision as a comparison of cost and gain, with the rational policymaker moved to adopt the initiative in question to the extent that the gains outweigh the costs. The nature and magnitude of costs are not considered in and of themselves; they take on meaning only in comparison to the expected gains. Such a theoretical formulation ignores the reality that *absolute* costs, when sufficiently great, are a very important consideration for policymakers. McGeorge Bundy, Special Assistant to the President for National Security Affairs in both the Kennedy and Johnson administrations, has criticized strategic analysts for their lack of realism in this regard:

> There is an enormous gulf between what political leaders really think about nuclear weapons and what is assumed in complex calculations of relative "advantage" in simulated strategic warfare. Think tank analysts can set levels of "acceptable" damage well up in the tens of millions of lives. They can assume that the loss of dozens of great cities is somehow a real choice for sane men. They are in an unreal world. In the real world of real political leaders—whether here or in the Soviet Union—a decision that would bring even one hydrogen bomb on one city of one's own country would be recognized in advance as a catastrophic blunder; ten bombs on ten cities would be a disaster beyond history; and a hundred bombs on a hundred cities are unthinkable. Yet this unthinkable level of human incineration is the least that could be expected by either side in response to any first strike in the next ten years, no matter what happens to weapons systems in the meantime.[15]

The absolute cost of nuclear war was probably an important restraining factor for American policymakers throughout the period of their nuclear superiority *vis à vis* the Soviet Union. No doubt, it continues to remain so. Consider the following hypothetical scenario. Soviet–American relations continue to deteriorate while America's political-military strength suffers a precipitous decline abroad following the overthrow of one or more important Third World client states and the further emasculation of NATO due to political differences among and within its member countries. Amidst these setbacks, the United States has exploited technological breakthroughs in laser beams and information processing that give it the capability to defend itself against a Soviet nuclear attack so effectively that only a couple of Soviet warheads at most could be expected to penetrate. However, the Soviets are known to be working frantically on perfecting their own defensive weapons, making it only a matter of time before they too possess an operational anti-missile system. In these circumstances some of the President's advisors might advise him to exploit America's strategic advantage by threatening or even starting a preventive war. It is very difficult to imagine a President agreeing to start such a war even if he was reasonably certain that the defensive systems would work as well as the Pentagon promised. In practice, of course,

15. McGeorge Bundy, "To Cap the Volcano," *Foreign Affairs*, Vol. 48, No. 1 (October 1969), pp. 1–20.

a prudent President with any experience of modern technology would fear, probably with good reason, that the defenses would break down just at the crucial moment, leaving the country open to destruction.

One important reason for the generally conceded reluctance of a President to start a nuclear war under even the most favorable conditions is the cost that it would still entail to the United States. Even a couple of warheads, if they fell on populated areas, would cause monumental loss of life, something probably sufficient in and of itself to discourage the President from seriously considering the attack option if he had any hope that war could be averted in the long term. As a general rule it may be that the more costly any contemplated foreign policy venture is judged to be, the more important absolute versus relative cost becomes in the minds of policymakers. If so, this calls into question the utility of a cost-gain calculus in predicting or explaining decisions with regard to the use of strategic nuclear weapons.

Moral and Psychological Costs

Third party estimates of the cost and gain of any foreign policy initiative are always based on tangible strategic and political considerations. The better analyses of this kind also include some consideration of the expected bureaucratic and domestic political effects of the policy in question. Even these assessments invariably ignore the moral and psychological dimension of decisions. However, these "hidden" costs and gains can be extremely important, even decisive, in affecting decisional outcomes. For reasons that the following analysis will make apparent, costs of this kind may be most pronounced in situations when policymakers must choose between war and peace.

Social psychologists describe policymakers as emotional beings rather than rational calculators.[16] In the words of Irving Janis and Leon Mann, "they are beset by doubts and uncertainties, struggle with incongruous longings, antipathies, and loyalties, and are reluctant to make irrevocable choices."[17] Important decisions generate internal conflict because policymakers are likely

---

16. Kurt Lewin, *Dynamic Theory of Personality* (New York: McGraw-Hill, 1935); Irving L. Janis, *Psychological Stress: Psychoanalytic and Behavioral Studies of Surgical Patients* (New York: John Wiley, 1958); Joseph H. de Rivera, *The Psychological Dimension of Foreign Policy* (Columbus, Ohio: Merrill, 1968); Irving L. Janis and Leon Mann, *Decision-Making: A Psychological Analysis of Conflict, Choice and Commitment* (New York: The Free Press, 1977).
17. Janis and Mann, *Decision-Making,* p. 15.

to experience opposing tendencies to accept and reject a given course of action. Decisional conflict and the stress it generates become acute when a policymaker realizes that there is a risk of serious loss associated with any course of action open to him. The stress can become crippling if this loss is perceived to entail the sacrifice of values that are extremely important to the policymaker. In these circumstances, he will be burdened with anticipatory feelings of shame, guilt, and related feelings of self-deprecation, which lower his self-esteem. A policymaker tends to cope with such situations by procrastinating, shifting responsibility for decision, and by "bolstering."[18] These affective responses detract from the quality of decision-making but are functional in the sense that they facilitate coping with stress. They may even be necessary for the policymaker to move confidently towards a decision.

It is difficult to conceive of a decision more fraught with stress than one to start a nuclear war. Mere contemplation of the act and its expected consequences—let alone its unexpected consequences—could be expected to arouse considerable anxiety on the part of almost any policymaker seriously considering it as an option. For this reason alone it is ludicrous to suppose that an American President or Soviet Premier could wake up one morning, decide that the correlation of forces was favorable, and so calmly give the order to push the button. Policymakers and people with any kind of ordinary feelings and emotions would find it difficult to assume responsibility for the death of untold millions of people and, should the war get out of hand, perhaps the destruction of society itself. Leaders who seriously contemplated the use of such destructive weapons would have a real need to do something in advance to reduce their anxiety and anticipatory guilt feelings before they could actually authorize a nuclear strike. There are a number of psychological techniques that could be employed towards this end.

Policymakers contemplating nuclear war could exaggerate the favorable consequences of their intended action. They could attempt to convince them-

---

18. Janis and Mann (ibid., pp. 74–95) use bolstering as an umbrella term that describes a number of psychological tactics designed to allow policymakers to entertain expectations of a successful outcome. Bolstering occurs when the policymaker has lost hope of finding an altogether satisfactory option and is unable to postpone a decision or foist the responsibility for it onto someone else. Instead, he commits himself to the least objectionable alternative and proceeds to exaggerate its positive consequences or minimize its negative ones. He may also deny the existence of his aversive feelings, emphasize the remoteness of the consequences, or attempt to minimize his personal responsibility for the decision once it is made. The policymaker continues to think about the problem but wards off anxiety by practicing selective attention and other forms of distorted information processing.

selves that the fruits of victory would be so great as to be worth almost any cost, much the way Hitler declared at the onset of World War II that the outcome of that conflict "would decide the future of the Reich for the next ten thousand years." In a present day context, Soviet or American leaders could try to convince those around them, but most of all themselves, that destruction of their adversary would not only remove the principal threat to their nation's survival but would permit them to organize an enduring world system based on their interests and ideological goals.

Policymakers contemplating nuclear war could also try to reduce whatever anxiety and guilt they felt by attempting to share responsibility for the decision. One way of doing this is to make the decision a group one and by doing so to spread at least psychologically the responsibility for it. Participating in a unanimous consensus along with respected fellow members of a congenial group will also bolster a policymaker's self-esteem.[19] Kennedy's "Ex Com" functioned in this manner. The solidarity it developed towards the end of its deliberations seems to have provided considerable emotional solace to the President, permitting him to confront more easily the series of difficult decisions that followed the imposition of the blockade.[20]

Policymakers might also try to evade or repudiate their responsibility for the decision. They could attempt to do this by attributing their decision to external pressures, thereby denying that they in any way personally want or approve of what they are about to do. They might succeed in convincing themselves, as did Bethmann Hollweg in 1914, that they were prisoners of circumstance, unwitting victims of the actions of adversaries or even allies.[21] As the German Chancellor told the Reichstag on August 4, 1914: "We have not willed war, it has been forced upon us."[22] Soviet and American leaders

19. For a provocative analysis of this phenomenon, see Irving L. Janis, *Groupthink*, 2nd. rev. ed. (Boston: Houghton, Mifflin, 1982).
20. Lebow, *Between Peace and War*, pp. 298–303, attempts to document this contention.
21. Ole R. Holsti, *Crisis, Escalation, War* (Montreal: McGill–Queen's University Press, 1972), pp. 22–32; Ithiel de Sola Pool and Allan Kessler, in "The Kaiser, the Tsar, and the Computer: Information Processing in a Crisis," *American Behavioral Scientist*, Vol. 8, No. 9 (1965), pp. 31–38, also emphasize the role of time pressure and information overload as stress-producing factors in the July crisis. The same point is made in a more general way by Dina A. Zinnes, Robert C. North, and Howard E. Koch, Jr., in "Capability, Threat and the Outbreak of War," in James Rosenau, ed., *International Politics and Foreign Policy: A Reader in Research and Theory* (New York: Free Press, 1961), pp. 469–482.
22. An English translation of this speech can be found in *Collected Diplomatic Documents Relating to the Outbreak of the European War* (London: His Majesty's Stationary Office, 1915), pp. 436–439.

could similarly attempt to portray themselves as boxed into a corner by the actions of the other superpower. If war appeared likely to develop out of a crisis, both sides might also seek refuge in the illusion, as did the major protagonists of World War I, that while they had no freedom of choice their adversary could still act in ways that would prevent war.

Germany in 1914

German policymaking in the years prior to 1914 offers some confirmation of the need of policymakers to resort to a variety of such psychological tactics when confronted with an anxiety-provoking decision. In 1905, 1909, and 1912 and again in 1914, members of the German General Staff urged political leaders to take advantage of Balkan crises to start a European war. Moltke himself declared to the Kaiser in December 1912 that war was inevitable and as for its timing, "the sooner the better."[23] The army feared that by 1917 Germany would soon lose its ability to win such a war because of Russian railway and military reforms. In light of the striking German fatalism about the inevitability of a European war, an attitude shared by political and military leaders alike from about 1905 on, it is important to inquire why Germany spurned war in the series of crises before 1914 but seemingly embraced it at that time. Historians of the period have offered several explanations for why this might have been so.

Fritz Fischer and others have argued that Germany's domestic contradictions had become more acute by 1914, compelling German leaders to embrace

23. For recent works stressing the eagerness of the general staff for war in 1914, see: Gerhard Ritter, *The Sword and the Scepter: The Problem of Militarism in Germany*, trans. Heinz Norden, 4 vols. (Coral Gables: University of Miami Press, 1969–1973), Vol. 2, pp. 227–275; Martin Kitchen, *The German Officer Corps, 1890–1914* (Oxford: Oxford University Press, 1968), pp. 96–114; Adolf Gasser, "Der deutsche Hegemonialkrieg von 1914," in Immanuel Geiss and Bernd Jürgen Wendt, eds., *Deutschland in der Weltpolitik des 19, und 20. Jahrhunderts: Festschrift für Fritz Fischer* (Düsseldorf: Bertelsmann Universitätsverlag, 1974), p. 310 ff.: K.H. Jarausch, *The Enigmatic Chancellor: Bethmann-Hollweg and the Hubris of Imperial Germany* (New Haven: Yale University Press, 1973), p. 181 ff.; Fritz Fischer, *War of Illusions: German Policies from 1911 to 1914*, trans. Marian Jackson (New York: Norton, 1975), is in agreement but stresses the commitment of the political authorities to provoking a showdown with the Entente from the very beginning of the crisis even if it led to war. For further documentation on Moltke's role in 1909, 1912, and 1914, see in addition Franz Conrad von Hötzendorff, *Aus Meiner Dienstzeit*, 5 vols. (Vienna: Rikola, 1921–1925), Vol. 1, p. 165 and Vol. 3, p. 670, documenting Moltke's calls for preventive war in 1909 and 1914; Walter Görlitz, *Der Kaiser . . . Aufzeichnungen des Chefs de Marinekabinetts Admiral Georg Alexander V. Müller, 1914–1918* (Göttingen: Munsterschmidtverlag, 1959), p. 124 ff. on 1912; Fischer, *War of Illusions*, pp. 161–164, 395–402, 453–455.

war as the only acceptable means of resolving them.[24] Luigi Albertini and, more recently, Volker Berghahn and Wolfgang Mommsen have stressed the failure of the German government to function effectively in 1914, allowing the "hardliners," especially the army, to gain the upper hand.[25] I have also emphasized the importance of faulty decision-making in bringing about war but traced this to complex political-psychological causes.[26] A recent and provocative article by David Kaiser attributes the differences to the Chancellor. Theobald von Bethmann Hollweg, Kaiser argues, was much more inclined to war than his predecessor, Bernard von Bülow, as he saw it as the

---

24. The thesis that Germany deliberately chose to exploit a window of opportunity in 1914 is most forcefully put by Fritz Fischer in two books: *Griff nach der Weltmacht* (Düsseldorf: Droste Verlag, 1961 [English translation: *Germany's Aims in the First World War* (New York: W.W. Norton, 1967)], and *Krieg der Illusionen* (Düsseldorf: Droste Verlag, 1969 [English translation: *War of Illusions* (New York: W.W. Norton, 1975)]. Fischer attributes Germany's decision for war to domestic problems. He argues that German leaders pursued an aggressive policy in order to overcome the stresses generated by the unstable political, economic, and social structure of the Reich. This interpretation has been developed further by Arno Mayer, "Domestic Causes of the First World War," in Leonard Krieger and Fritz Stern, eds., *The Responsibility of Power* (Garden City, N.Y.: Doubleday, 1967), pp. 308–324; Hans-Ulrich Wehler, *Der Deutsche Kaiserreich, 1871–1918* (Göttingen: Vandenhoeck and Ruprecht, 1977). Fischer's student, Imanuel Geiss, *German Foreign Policy, 1871–1914* (London: Routledge and Kegan Paul, 1976), argues that Sarajevo was "hardly more than the cue for the Reich to rush into action"; Volker R. Berghahn, *Germany and the Approach of War in 1914* (New York: St. Martin's Press, 1973), offers a more balanced approach. The Fischer thesis has been contested by other German historians, among them: Gerhard Ritter, *Staatskunst und Kriegshandwerk: Das Problem des Militarismus in Deutschland*, 2nd ed. rev. (Munich: R. Oldenbourg Verlag, 1965), Vol. 2; Wolfgang Mommsen, "Domestic Factors in German Foreign Policy Before 1914," *Journal of Central European History*, Vol. 6 (March 1973), pp. 11–43, who argues that war broke out because the German government did not function effectively in the July crisis; Geoff Eley, *Reshaping the German Right: Radical Nationalism and Political Change After Bismarck* (New Haven: Yale University Press, 1980), who disputes the extent to which the German government sought to manipulate nationalism in its foreign policy; and Richard Ned Lebow, *Between Peace and War: The Nature of International Crisis* (Baltimore: Johns Hopkins University Press, 1981), pp. 101–147, who attributes war to faulty decision-making on the part of German leaders and their foreign policy bureaucracy. David E. Kaiser, "Germany and the Origins of the First World War," *Journal of Modern History*, Vol. 55 (September 1983), pp. 442–474, makes a strong case that the literature from Fischer on has grossly exaggerated the extent to which German foreign policy before 1914 was designed to achieve domestic goals.

25. Luigi Albertini, *The Origins of the War of 1914*, trans. and ed. Isabella M. Massey, 3 vols. (Oxford: Oxford University Press, 1962); Volker R. Berghahn, *Germany and the Approach of War in 1914* (New York: St. Martin's Press, 1973); Wolfgang Mommsen, "Domestic Factors in German Foreign Policy Before 1914," *Journal of Central European History*, Vol. 6 (March 1973), pp. 11–43; and Michael R. Gordon, "Domestic Conflict and the Origins of the First World War: The British and German Cases," *Journal of Modern History*, Vol. 46 (June 1974), pp. 191–226, which in many ways is the most sophisticated presentation of the argument that domestic problems both influenced German foreign policy and degraded its implementation.

26. Lebow, *Between Peace and War*, pp. 101–147.

only means of escaping encirclement and of achieving Germany's rightful place among the powers.[27]

None of these explanations is entirely satisfying. The thesis of domestic causation in particular has recently come under strong attack for failing to demonstrate just how domestic concerns actually influenced specific policies.[28] The governmental breakdown thesis suffers from the weakness that most of its proponents fail to offer convincing reasons for why the German government should have performed so poorly in the crisis. Kaiser in turn can be criticized for overstating the differences between the two chancellors. He also fails to explain why Bethmann Hollweg rejected the army's request for war in 1912 but apparently acceded to it less than two years later. Whatever the partial merit of these several explanations, it is worth considering yet another alternative: that the important difference between 1914 and earlier crises was its very different decisional context. This should not be considered a complete explanation by itself but rather as an important missing piece of the puzzle of why 1914 was different from the crises that preceded it.

In the series of pre-1914 crises, war could only have come about as the result of a conscious and relatively unambiguous decision by German leaders to draw their sword. In 1905, neither France nor Britain was disposed towards war and the French for their part were doing everything in their power to avoid giving Berlin a pretext for military action. In 1909, Russia's capitulation to Germany's ultimatum, something Chancellor Bülow and the Kaiser expected, removed any but the most transparent excuse for war. In 1911 and 1912, Germany was a peripheral actor in a complex Balkan drama and, given Austria's indecision, would have had to have gone to some length to engineer an appropriate *casus belli.* Kiderlen-Wächter, foreign minister during the 1911 Agadir crisis, lamented, "If we do not conjure up a war into being, no one else certainly will do so."[29]

War in any of these situations would have required German leaders to have imposed their will on a course of events that in each case appeared to

---

27. Kaiser, "Germany and the Origins of the First World War." pp. 442–474.
28. In addition to the Mommsen and Kaiser articles already cited, see Geoff Ely, *Reshaping the German Right;* and Isabel V. Hull, *The Entourage of Kaiser Wilhelm II, 1888–1918* (New York: Cambridge University Press, 1982), who rightly criticizes the tendency among German historians to explain events in terms of abstract social forces without ever explaining how these forces actually influenced policy in practice.
29. Hermann Kantorowicz, *The Spirit of British Policy and the Myth of the Encirclement of Germany* (London, 1932), p. 360.

be heading towards a peaceful outcome. German leaders could not easily have claimed either at the time or afterwards that they had been prisoners of circumstances. The choice for peace or war lay in their hands, they knew it, and so did the other powers. Responsibility for war would also have been difficult to evade because of the diplomatic scenario Germany would have had to devise and implement in each case in order to bring about a war. In these circumstances, as events made clear, Germany's leaders were unprepared to start a war.

In 1905, the documents indicate that Kaiser and Chancellor alike were relieved that the peace had been preserved in spite of their belief that war was inevitable sooner or later and their realization that time worked to Germany's disadvantage. Both men were actually pacific by inclination. The Kaiser especially was known to fear the prospect of war. By most accounts, his sword-rattling and blustery public demeanor were for public consumption. Admiral Tirpitz observed: "When the Emperor did not consider the peace to be threatened he liked to give full play to his reminiscences of famous ancestors"; but "in moments which he realized to be critical he proceeded with extraordinary caution."[30] Bülow, Chancellor during both the Moroccan and Bosnian crises, also dismissed Wilhelm's aggressiveness as a façade. He thought it was designed to compensate for his pronounced feelings of inadequacy.[31] The Kaiser's subsequent withdrawal from responsibility and reality during the war years, 1914–1918, certainly confirms this portrayal of him as a man incapable of coping with the stress that conflict engendered.[32]

Bethmann Hollweg, Bülow's successor as Chancellor, is a more puzzling and controversial figure.[33] He was prescient enough to realize that the do-

30. Alfred von Tirpitz, *Politische Dokumente*, 2 vols. (Berlin: Cotta, 1924–1926), Vol. 1, p. 242; see also Michael Balfour, *The Kaiser and His Times* (Boston: Houghton, Mifflin, 1964); Hull, *The Entourage of Kaiser Wilhelm II*, pp. 236–265.

31. Bernhard von Bülow, *Memoirs*, trans. F.A. Voight, 4 vols. (Boston: Little, Brown, 1931), Vol 3, p. 149.

32. For a description of Wilhelm's withdrawal from reality during the war, see Alastair Horne, *The Price of Glory: Verdun 1916* (Harmondsworth: Penguin Books, 1964), pp. 45–46.

33. The traditional literature treats Bethmann Hollweg sympathetically, as someone who did not desire war but proved incapable of preventing it. Konrad H. Jarausch, in "The Illusion of Limited War: Chancellor Bethmann-Hollweg's Calculated Risk, July 1914," *Journal of Central European History*, Vol. 2 (March 1969), pp. 48–76; and Fritz Stern, in "Bethmann-Hollweg and the War: The Limits of Responsibility," in Leonard Kreiger and Fritz Stern, eds., *The Responsibility of Power: Historical Essays in Honor of Hajo Holborn* (Garden City, N.Y.: Doubleday, 1969), pp. 271–307, attempt to reconstruct the Chancellor's attitudes and objectives during the crisis on the basis of the diary entries of Kurt Riezler, his long-time political confidant and secretary. Jarausch argues that Bethmann Hollweg risked a general war in the hope and expectation of

mestic repercussions of a European war, especially a protracted one, could be just the reverse of that anticipated by the conservatives. "A world war," he warned, "would greatly increase the power of social democracy because it had preached peace and would bring down many a throne."[34] At the same time, he was increasingly obsessed by Russia's growing military strength and how it would soon "overwhelm" Germany. "In a few years," he confided to his private secretary in July 1914, "Russia will be supreme and Germany her first, lonely victim."[35] The Chancellor was a man torn by opposing tendencies, a tension that was outwardly manifested as indecision. To his contemporaries he thus appeared something of a fatalist—one who, in spite of his reservations about the wisdom of war, nevertheless felt powerless to counteract the prevalent view that war was necessary.[36] At the same time, he was also unprepared to act in response to the pleadings of the generals and actually provoke such a conflict.

The picture of Chancellor and Kaiser that emerges is that of rather ordinary men, a bit more insecure than most in the case of Wilhelm. They were caught on the horns of a painful dilemma: Germany's vital interests seemingly required war but their personality structures would not permit them to accept responsibility for such an awesome venture. Up until 1914, their response was procrastination, a hallmark of defensive avoidance. By deferring a decision that was too difficult for them to make, German leaders preserved their psychological equilibrium. The July crisis at last offered them an escape

---

breaking up the Entente and bringing about a new alignment more favorable to Germany. Fritz Stern, in fundamental agreement with Jarausch, states that "The Riezler diary sustains the view that Bethmann in early July had resolved on a forward course; by means of forceful diplomacy and a local Austrian war against Serbia he intended to detach England or Russia from the Entente or—if that failed—to risk a general war over an opportune issue at a still opportune moment." Kaiser, "Germany and the Origins of the First World War," makes the same argument. More general interpretations of Bethmann Hollweg's policy are to be found in Karl Dietrich Erdmann, "Zur Beurteilung Bethmann Hollwegs," *Geschichte in Wissenschaft und Unterricht*, Vol. 15 (September 1964), pp. 525–540, and Andreas Hillgruber, "Riezler's Theorie des kalkulierten Risikos and Bethmann-Hollwegs politische Konzeption in der Julikrise 1914," *Historische Zeitschrift*, Vol. 202 (April 1966), pp. 333–351. The latter analyzes the Chancellor's crisis policy in terms of Riezler's prewar writings.

34. Cited by Jarausch in *Enigmatic Chancellor*, p. 58.

35. See Fritz Stern, "Bethmann-Hollweg and the War," pp. 271–307; Egmont Zechlin, "Deutschland zwischen Kabinettskrieg und Wirtschaftkrieg, Politik und Kriegführung in den ersten Monaten des Weltkrieges 1914," *Historische Zeitschrift*, Vol. 199 (1964), pp. 347–352; Kaiser, "Germany and the Origins of the First World War," pp. 442–474.

36. Jarausch, *Enigmatic Chancellor*, pp. 148–184; Fritz Stern "Bethmann-Hollweg and the War," pp. 271–307; Lebow, *Between Peace and War*, pp. 254–259, on the German belief in the inevitability of war in the years before 1914.

from the decisional dilemma that had hung over their heads for almost a decade.

The crisis was precipitated by the assassination of Archduke Franz Ferdinand by a Serbian nationalist. This was seized upon by Leopold Berchtold, the Austrian Chancellor, and Franz Conrad von Hötzendorf, chief of staff, as a pretext for destroying Serbia. But Emperor Franz Josef and Peter Tisza, the Hungarian leader, were disinclined towards war. Berchtold accordingly dispatched Count Hoyos to Berlin in an attempt to secure German backing for war in the hope that this would force a consensus for action in Vienna.[37]

When Hoyos met the German Kaiser and the Chancellor in Potsdam, he did not appear to confront them with a choice between war or peace but only a request to back Austria should Russia contemplate intervention in support of Serbia. German leaders did not consider this eventuality likely to arise as they believed that Russia would back down as it had in 1909 when confronted with the prospect of war with Germany and Austria. If Russia unexpectedly intervened in Serbia's defense, Germany's political leaders and diplomats refused to believe that France would come to Russia's assistance, nor Britain to France's if the conflict spread to the West. When Chancellor Bethmann Hollweg and the Kaiser voiced their support of Austria and gave it the so-called "blank check," war seemed remote and a full-scale European war even more so.[38]

As we now know, German support for Austria set in motion a chain of events that led to a world war: it precipitated Austria's mobilization and declaration of war against Serbia, Russia's subsequent mobilization against Austria and, of necessity, against Germany as well. Russian mobilization provoked German ultimatums to Russia, Belgium, and France and their rejection prompted German mobilization, which was the virtual equivalent of war. When Berlin confronted Russian mobilization or, more accurately, the premature and exaggerated reports of Russian mobilization flooding in through the channels of German military intelligence, German leaders convinced themselves that they were only reacting to Russian initiatives; St. Petersburg, not they, bore the brunt of responsibility for the war that was about to start because Russia, they believed, had secretly refused steadfast-

37. For a good secondary account of Austria's role in the July crisis, see Ritter, *The Sword and the Scepter,* Vol. 2, pp. 227–263.
38. On German expectations that a Balkan War could be kept limited, see Lebow, *Between Peace and War,* pp. 119–135.

edly to halt its military preparations. German leaders were convinced that they had no choice but to mobilize in return.

What had begun as a diplomatic offensive passed beyond the bounds of politics because German political leaders possessed neither the courage nor good sense to alter their policy in mid-crisis. Inescapably confronted with the fact that an Austro–Serbian war could not be localized, the Chancellor and the Kaiser were overcome by anxiety. Neither man was fully willing to admit the probable outcome of continued support of Austrian bellicosity nor prepared to accept the responsibility for a radical reorientation of German policy. Their actions from the fateful night of July 29–30 to the outbreak of war betrayed irresolution, bewilderment, and loss of self-confidence. The Kaiser oscillated between moods of profound optimism and despair. His hypervigilant Chancellor vacillated between the very extremes of available policy options. At first appalled by the thought of a European war, he sought to restrain Austria. Later, influenced by Moltke, he again urged military action upon Vienna. Both Kaiser and Chancellor ultimately lapsed into passive acceptance of the inevitability of war although they continued to clutch at the hope of British neutrality.[39]

Wilhelm and Bethmann Hollweg's abdication of responsibility allowed Moltke to make the crucial decisions. The strange behavior of the two men seems best interpreted as an effort to protect their own personality structures by shifting responsibility for the decision to mobilize onto Moltke's shoulders. Their last-minute intervention in the crisis, which consisted of frenzied diplomatic efforts to cast the blame for war onto Russia, was probably motivated by the same goal. Moltke, the one German leader who outwardly displayed clarity of purpose throughout the crisis, also seems to have suffered deeply. Later events suggest that his *sangfroid* was quite deceptive. Moltke suffered acute stress during the crisis but forced himself to carry his policy forward. Once war actually broke out, he began a serious psychological decline that culminated in a nervous breakdown.[40]

---

39. Lebow, *Between Peace and War*, pp. 135–147.
40. Moltke's incapacitation, so disastrous for the course of the campaign in France, is usually attributed to his disturbing confrontation with the Kaiser on August 1 when he was rebuked by Wilhelm for adamantly refusing to consider redirecting the German offensive against Russia instead of France in response to information, erroneous as it turned out, that France and Britain would remain neutral. It is difficult to believe that this exchange, unpleasant as it no doubt was, could in and of itself have been responsible for Moltke's subsequent breakdown. Nevertheless, those around him later marked his decline from this incident. The episode becomes more readily understandable if Moltke is seen as a man who was already suffering from pronounced guilt

The peculiar structure of the July crisis permitted German leaders to move towards war by a series of incremental and deceptive decisions. The first, and perhaps the most important, was the decision to back Austria. This could be rationalized as carrying only a remote risk of war. Later, the decision to mobilize, a *de facto* declaration of war, could be described by German leaders to themselves as unavoidable; they portrayed themselves at that time and later as men without choices who had becme trapped by the actions of others.

The decisions that led to war in 1914 were not only cumulative but international. The crisis scenario the Germans devised or stumbled into, depending upon one's interpretation, compelled other actors, most notably Austria, Serbia, Russia, and France, to share responsibility for the war. This made it easier for Germany's leaders to cope with whatever guilt their behavior aroused. Finally, there was the relative passivity of the Kaiser and the Chancellor at the height of the crisis. In past crises, German inaction had facilitated peaceful resolution; as nobody wanted war, the tensions wound down. On this occasion, however, it made war more likely given the momentum of events. German policy came to resemble a stone rolling downhill; it gathered speed along a course that was neither stopped nor altered by those who had set it in motion.[41]

In retrospect, the remarkable feature of 1914 is not that Germany went to war but rather the tremendous psychological difficulty German leaders had in doing so. The crisis was extremely anxiety-provoking in spite of a decision-making context that facilitated the use of all kinds of techniques of defensive

---

feelings because of his leading role in pushing the crisis towards war. What Moltke desperately needed was support and reassurance. Criticism and rejection were devastating to him given his vulnerabilities at the moment and his already dangerously low sense of self-esteem.

41. The illusion that the war would be short and glorious, so widespread among the participants and so baffling to the historians, may also have served an important anxiety-reducing function. Certainly, there was ample evidence from the American Civil and the Russo–Japanese wars that the concentrated five protective earthworks and belts of barbed wire gave a tremendous advantage to the well-prepared defense. The fact that this evidence was so often overlooked in favor of the experience of the Franco–Prussian War, a conflict whose lessons in this regard were in any case ambiguous, suggested the existence of motivated bias on the part of European military leaderships. This bias probably derived from their commitment to the idea of the offensive. Political leaders, worried about the cost and consequences of a lengthy war of attrition, would have had an incentive to believe in the viability of the offensive. For German thoughts on this subject, see Lancelot L. Farrar, Jr., *The Short War: German Policy, Strategy, and Domestic Affairs, August–December 1914* (Santa Barbara, Calif.: ABC-CLIO, 1973). Jack L. Snyder, *The Ideology of the Offensive* (Ithaca, N.Y.: Cornell University Press, 1984), offers a provocative comparative analysis of the European military bias in favor of the offensive and the contribution it made to war in 1914.

avoidance. Given the unwillingness of German leaders to accept the responsibility for starting a war, it is certainly difficult to imagine how Kaiser, Chancellor, Foreign Office, and even military establishment could have screwed up their courage to go to war in a decisional context that would have compelled them to accept more openly their share of responsibility for it. If the Archduke had not been assassinated in 1914, giving rise to the unusual opportunity I have just described, it seems quite likely that Germany would have reached that fateful year of 1917 still at peace with its neighbors. If so, its leaders might have discovered that their fears of a window of vulnerability were greatly exaggerated—that their adversaries were constrained from attacking Germany for many of the same reasons that had prevented Germany from exploiting its window of opportunity in the years before 1914.

The United States: 1948–1960

There are many structural similarities between the German situation before 1914 and the American one in the 1950s. Like the Germans, the Americans possessed a clear-cut strategic advantage but one they realized would diminish with every passing year until victory became an unattainable goal. Again like the Germans, many American officials thought that war between themselves and their adversary was all but unavoidable. NSC-68, the most important foreign policy document of the decade, predicted that the Soviet Union would attack the United States by 1955 at the latest unless the Truman Administration initiated a far-reaching program to maintain American nuclear superiority.[42]

Germany's security predicament in the decade before 1914 had led a number of prominent Germans to call for a preventive war against France and Russia. Some Americans after World War II urged their government to strike against the Soviet Union. An early voice in this regard was Dr. Virgil Jordan, president of the National Industrial Conference Board. In a speech given in February 1946 to the Union League Club of Philadelphia he told his audience of five hundred, among them many prominent industrialists, that the United States must subdue the Soviet Union, "a primitive, impoverished, predatory Asiatic despotism," or be destroyed by it. Moscow was to be given an

42. *Foreign Relations of the United States, 1950*, Vol. 1, pp. 235–292.

ultimatum to disarm and to submit to American control over all its industrial processes. Rejection of this ultimatum would provide the pretext for unleashing atomic war.[43]

An even more outspoken advocate of preventive war was Major George Fielding Eliot, former military correspondent for the *New York Herald Tribune.* In a book published in 1949, and quite inappropriately titled *If Russia Strikes,* he called for an atomic onslaught against the Soviet Union before it developed atomic bombs of its own. Once this happened, Eliot warned, the West, especially Western Europe, would become a hostage to be destroyed in retaliation for any American attack. "The only way to prevent or mitigate such a massacre," he declared, "would be to strike quick[ly] and hard at the centers of Soviet power, and so shatter the will and smash the strength of the Soviet monster that his reactions against helpless people will be no more than dying convulsions."[44]

The notion of preventive war also appears to have received some serious attention in official circles. Winston Churchill urged an Anglo–American nuclear attack against the Soviet Union at the time of the Berlin blockade, a suggestion seconded by American Secretary of Defense Louis Johnson. Churchill's proposal was rejected out of hand by Prime Minister Clement Atlee and President Truman.[45] Preventive war was also discussed by members of the Joint Committee on Atomic Energy in the immediate aftermath of the first Soviet atomic test in August 1949[46] It was among the four "Possible Courses of Action" considered by Paul Nitze's Policy Planning Staff in NSC-68, received by President Truman on April 11, 1950.[47] In August of that year,

---

43. Jordan's speech was published as a pamphlet and in *Vital Speeches* for May 1, 1946. Cited in D.F. Fleming, *The Cold War and Its Origins, 1917–1960* (Garden City, N.Y.: Doubleday, 1961), pp. 392–394. Bertrand Russell issued his infamous call for a nuclear ultimatum a few months later, in October 1946. He called for the Western democracies to pressure Moscow into accepting a world government: "The only possible way" of doing this would be by "a mixture of cajolery and threat, making it plain to the Soviet authorities that refusal will entail disaster, while acceptance will not." "The Prevention of War," in Morton Grodzins and Eugene Rabinowitch, eds., *The Atomic Age: Scientists in National and World Affairs* (New York: Basic Books, 1963), pp. 100–106.

44. George Fielding Eliot, *If Russia Strikes* (Indianapolis: Bobbs-Merril, 1949); Fleming, *Cold War and Its Origins,* pp. 395–397.

45. *The Washington Post,* January 3, 1979, p. 12, citing newly released British cabinet papers; Harry S Truman, *Memoirs,* 2 vols. (Garden City, N.Y.: Doubleday, 1955–1956), Vol. 2, pp. 355–383.

46. Herken, *Winning Weapon,* p. 318.

47. *Foreign Relations of the United States,* 1950, pp. 276–281; Samuel F. Wells, Jr., "Sounding the Tocsin: NSC-68 and the Soviet Threat," *International Security,* Vol. 4, No. 2 (Fall 1979), pp. 116–158; Herken, *Winning Weapon,* p. 318, citing records of the Joint Chiefs of Staff.

Secretary of the Navy Francis P. Matthews openly called for a preventive war against the Soviet Union. Americans, as he put it in an early example of Pentagon doublespeak, would become the first "aggressors for peace."[48] Senior Air Force officers continued to talk about preventive war as late as 1953.[49]

Although bruited about as a policy option, preventive war was repudiated by President Truman as a policy "unthinkable for rational men."[50] Truman even fired Secretary Matthews for having broached the subject publicly.[51] Later presidents also spurned all suggestions of preventive war. Why did the United States, unlike Germany, fail to exploit its diminishing strategic superiority?

One important difference between the two situations was that American leaders were never as convinced as their German counterparts that a preventive war would succeed. In 1950, even the Air Force demurred from claiming that an atomic blitz would in and of itself succeed in defeating the Soviet Union. The Army absolutely ridiculed the notion. General Omar Bradley repeatedly warned that the notion of a short atomic war was "folly." He and other Army officers argued that atomic bombs dropped on Soviet cities would do nothing to halt the Soviet offensive in Western Europe that would surely follow hard on the heels of any American attack. Victory, Bradley was convinced, would require an "extended, bloody and horrible" struggle in Europe, reminiscent of World War II.[52] The authors of NSC-68 reached the same conclusion.[53] The Air Force's own study of strategic bombing in World War II, which came out about this time, also helped to raise doubts about the efficacy of a preventive war. It found that three years of extensive allied bombing of Germany had done little to damage German morale or industrial

---

48. Matthews' speech is cited in Alfred Vagts, *Defense and Diplomacy* (New York: King's Crown, 1956), pp. 329–333.
49. Curtis LeMay with McKinley Kantor, *Mission With LeMay* (Garden City, N.Y.: Doubleday, 1965), p. 481.
50. "The President's Farewell Address to the American People, January 15, 1953," *Public Papers of the Presidents: Harry S. Truman, 1952–53* (Washington, D.C.: U.S. Government Printing Office, 1979), pp. 1197–1202.
51. Truman, *Memoirs,* Vol. 2, p. 383; Bernard Brodie, *Strategy in the Missile Age* (Princeton: Princeton University Press, 1965), pp. 229–232; Vagts, *Defense and Diplomacy*, pp. 329–335; George Quester, *Nuclear Diplomacy: The First Twenty-Five Years* (Cambridge: Dunellen, 1970), pp. 67–69; LeMay, *Mission With LeMay*, pp. 481–482.
52. Omar Bradley, "This Way Lies Peace," *Saturday Evening Post*, October 15, 1949, and his speech before the Chicago Economic Club as reported in *The New York Times*, October 29, 1948. Both references cited in Schilling, "Politics of National Defense," pp. 166–167.
53. See footnote 47.

capacity. There seemed little reason to believe that an atomic air offensive against the Soviet Union, which would inflict less overall destruction because of the limited number of weapons available at the time, would be any more successful in achieving these goals.[54] The Harmon Report, which reviewed the role of atomic bombs in American strategy, concluded that they could also be counterproductive in a political sense: "For the majority of Soviet people, atomic bombing would validate Soviet propaganda against foreign powers, stimulate resentment against the United States, unify these people and increase their will to fight."[55]

Doubt about the feasibility of preventive war is sometimes cited as the principal consideration inhibiting policymakers in the late 1940s and early 1950s from pursuing this option.[56] The importance of technical problems should not be discounted, but neither should the moral and psychological impediments to preventive war. Many political and military leaders were disturbed by the prospect of using atomic weapons, even in retaliation for a Soviet attack in Western Europe. Dwight Eisenhower, chief of staff of the Army in 1949, publicly expressed his indignation at the thought that his country would vouchsafe its security to "a weapon that might destroy millions overnight."[57] Omar Bradley, Eisenhower's successor, also attacked the immorality of an atomic blitz. He exclaimed to reporters in July 1949: "Ours is a world of nuclear giants and ethical infants. We know more about war than we know about peace, more about killing than about living."[58]

---

54. Hanson W. Baldwin in *The New York Times*, August 8, 1949, made public the most important findings of the U.S. Strategic Bombing Survey. David MacIsaacs, *Strategic Bombing in World War II: The Story of the United States Strategic Bombing Survey* (New York: Garland, 1976), provides a useful history of the Survey. For a discussion of the World War II experience and its impact upon the development of early postwar nuclear strategy, see Brodie, chapters 3 and 4.
55. The Harmon Report, *Evaluation of Effect on Soviet War Effort Resulting from the Strategic Air Offensive*, May 11, 1949. Reprinted in Thomas H. Etzold and John Lewis Gaddis, eds., *Containment: Documents on American Policy and Strategy, 1945–1950* (New York: Columbia University Press, 1978), pp. 360–364. *History of the Joint Chiefs of Staff*, pp. 311–315.
56. See, for example, Quester, *Nuclear Diplomacy*, pp. 1–8, 67–68; Fleming, *Cold War and Its Origins*, pp. 391–415; Herken, *Winning Weapon*, p. 318, who also argues that preventive war "before the Russians had an atomic bomb would come too soon, but one made once they had broken the atomic monopoly would be too late." This is unconvincing as a Soviet test was not the equivalent of possessing an arsenal of weapons and the means to deliver them. Presumably, a preventive war could have been waged in the fall of 1949 with little or no fear of Soviet atomic retaliation.
57. *Daily Compass* (New York), July 1, 1949. Quoted in Fleming, *Cold War and Its Origins*, pp. 397–398.
58. Ibid.

The Joint Chiefs of Staff as a whole refused to believe that the American people would sanction an unprovoked atomic attack against the Soviet Union.[59] So did the War Department.[60] Secretary of State George Marshall expressed this opinion in March 1948 to the Senate Armed Services Committee, at least one of whose members had voiced support for preventive war. He reminded the senators that atom bombing Russian cities meant the wholescale killing of children and other civilians. The United States, Marshall argued, had conducted strategic bombing during World War II because the prior actions of the Axis powers had so infuriated the American people:

"But it was a terrible thing to have to use that type of power. If you are confronted with the use of that type of power in the beginning of war you are also confronted with a very certain reaction of the American people. They have to be driven very hard before they will agree to such a drastic use of force."[61]

Secretary of Defense Forrestal, among others, doubted that the American people were so inhibited and President Truman himself declared in September 1948 that if he had to use the bomb he would not shrink from doing so.[62] Truman, however, had in mind a situation in which the United States was *responding* to a Soviet attack, either conventional or nuclear, not initiating it.

---

59. Herken, *Winning Weapon*, p. 318, citing the records of the Joint Chiefs of Staff.
60. A 1947 paper, "The Effects of the Atomic Bomb on National Security," an expression of War Department thinking on this subject, declared: "We are prevented by our form of Government and our constitutional processes from launching surprise attacks against potential enemies." Quoted in Bernard Brodie and Eilene Galloway, *The Atomic Bomb and the Armed Services*, Public Affairs Bulletin No. 55 (Washington, D.C.: Library of Congress Legislative Reference Service, 1947), p. 70.
61. Schilling, "Politics of National Defense," p. 173, citing Marshall's testimony before the Senate Armed Services Committee Hearings on Universal Military Training in March 1948.
62. Walter Millis, ed., *The Forrestal Diaries* (New York: Viking Press, 1951), pp. 461–462, 486–488. Forrestal sounded out officials on both sides of the Atlantic about wartime use of atomic weapons: "Marshall was to quote to him a remark of John Foster Dulles that 'the American people would execute you if you did not use the bomb in the event of war'; Clay said that he 'would not hesitate to use the atomic bomb and would hit Moscow and Leningrad first'; Winston Churchill, going even further, told him that the United States had erred in minimizing the destructive power of the weapon—to do so was to lend dangerous encouragement to the Russians" (pp. 433–434, 457). Truman was different from other Presidents as he had actually authorized the use of atomic weapons against the Japanese. Whatever his understanding of the destructive potential of these weapons at the time—which does not seem to have been profound—he had a political and presumably psychological need to justify his action in later years. His statements about atomic and nuclear weapons both in and out of office must be interpreted in this light. It may explain why he made tough public statements about his willingness to use such weapons. Privately, as we have seen, he was exceedingly reluctant to contemplate their use.

For him it was clearly a weapon of last resort. Truman confided to David Lilienthal:

"I don't think we ought to use this thing unless we absolutely have to. It is a terrible thing to order the use of something that is so terribly destructive beyond anything we have ever had. You have got to understand that this isn't a military weapon. It is used to wipe out women, children and unarmed people, and not for military use. So we have to treat this differently from rifles and cannon and ordinary things like that."[63]

As President, Eisenhower retained his distaste for the indiscriminate use of nuclear weapons against civilian populations although he contemplated employing such weapons on three occasions.[64] He also authorized the Joint Chiefs to plan for nuclear retaliation in the advent of a Soviet attack on the United States or its armed forces.[65] In May 1954 Eisenhower nevertheless rejected a recommendation for the Advance Study Group of the Joint Chiefs of Staff that the United States consider "deliberately precipitating war with the USSR in the near future" before Soviet thermonuclear capability posed a "real menace."[66] Eisenhower was also reluctant to consider preemption as a

---

63. David Lilienthal, *The Journals of David E. Lilienthal*, Volume 2, *The Atomic Energy Years, 1945–1950* (New York: Harper and Row, 1964), p. 391.
64. The three occasions were: 1953, when the administration considered the use of nuclear weapons against China as a means of ending the Korean War; and 1955 and 1958, when China again became a possible target during crises over the offshore islands of Quemoy and Matsu. Dwight David Eisenhower, *Mandate for Change: The White House Years, 1953–1956* (Garden City, N.Y.: Doubleday, 1963), pp. 180, 476; Morton H. Halperin, *The 1958 Taiwan Straits Crisis: A Documented History* (Santa Monica, Calif.: Rand Corporation, 1966), pp. v–xiii; Desmond Ball, "U.S. Strategic Forces: How Would They Be Used?," *International Security*, Vol. 7, No. 3 (Winter 1982–1983), pp. 31–60. On at least one occasion Eisenhower also apparently thought about preventive war, albeit in the most abstract kind of way. He wrote to John Foster Dulles in September 1953 that the United States might someday find itself vulnerable to a Soviet first strike. Its security would thereafter depend upon its capability to hit back even harder: "But if the contest to maintain this relative position should have to continue indefinitely, the cost would either drive us to war—or into some form of dictatorial government. In such circumstances, we would be forced to consider whether or not our duty to future generations did not require us to initiate war at the most propitious moment we could designate." Memorandum, Eisenhower to Dulles, September 8, 1953, DDE Diary, August–September 1953, Folder 2, DDE Diary, Box 3, ACWF-EPP, DDEL. Cited in Rosenberg, "The Origins of Overkill," p. 33.
65. David Alan Rosenberg, "'A Smoking Radiating Ruin at the End of Two Hours': Documents on American Plans for Nuclear War with the Soviet Union, 1954–55," *International Security*, Vol. 6, No. 3 (Winter 1981–82), pp. 3–38, documents Eisenhower's decisions from papers of the Joint Chief of Staff.
66. NSC 5440/1, December 28, 1954; approved as Basic National Security Policy in NSC 5501, January 6, 1955, NSC-MMB, Paragraph 35. See also M.B. Ridgway, Memo for the Special Assistant to the President for National Security Affairs, Subject: Review of Basic National Security Policy, 22 November 1954, Historical Record Folder, Box 30, Ridgway Papers, cited in Rosenberg, "The Origins of Overkill," p. 34.

policy option even though the Joint Chiefs advised him that the time was fast approaching when the Strategic Air Command could deliver a "knock out blow" against the Soviet Union. In January 1956, Eisenhower noted in his diary that there were compelling moral as well as constitutional obstacles to launching a surprise attack. "It would not be only against our traditions," he wrote, "but it would appear to be impossible unless the Congress would meet in a highly secret session and vote a declaration of war."[67] It was "impossible," Eisenhower thought, that any such thing would ever occur.

The Kennedy Administration is reported to have toyed with the possibility of carrying out a limited attack against the Chinese nuclear program.[68] But when push came to shove in the Cuban missile crisis, Kennedy decided against the airstrike option at least in part because he envisaged it as a Pearl Harbor in reverse with himself cast in the role of Tojo.[69] If the moral cost of a "sneak attack" with conventional weapons against Soviet missiles in Cuba was too much for President Kennedy, it is difficult to conceive of the circumstances in which he would have assumed the responsibility for a nuclear "sneak attack."

It may be that the overwhelming sense of guilt and shame that postwar presidents knew they would have felt in the aftermaths of pushing the button was another important reason why preventive war was rejected by every administration during the years of unquestioned American nuclear superiority. In this connection it must be remembered that a nuclear assault upon the Soviet Union in the late 1950s or early 1960s, when such an attack was finally deemed capable of destroying that country as a functioning state, would have been awesome in scale. It would have resulted in many more casualties than most contemporary attack options because of the city-busting

---

67. Ibid., citing the declassified version of NSC 5602/1 of March 15, 1956.
68. Ball, "U.S. Strategic Forces: How Would They Be Used?," p. 43. No reference is given.
69. Elie Abel, *The Missile Crisis* (Philadelphia: Lippincott, 1966), p. 64. On October 13, 1964, in a campaign speech, Robert Kennedy reported that American intelligence had estimated that twenty-five thousand Cubans would be killed in a "surgical" airstrike against Cuba. "We could have gone in and knocked out all their bases—there wasn't any question about it—and then started bargaining." But, Kennedy insisted the President would have no part of a "Pearl Harbor in reverse," because of "his belief in what is right and what is wrong." In *Thirteen Days: A Memoir of the Cuban Missile Crisis* (New York: Norton, 1969), p. 31, Kennedy related that he passed a note to the President when the airstrike was being discussed that read: "I now know how Tojo felt when he was planning Pearl Harbor." Dean Acheson, in "Homage to Plain Dumb Luck," *Esquire*, February 1969, p. 76, reports that it was Robert Kennedy who made the analogy and that he, Acheson, challenged it as "thoroughly false and pejorative." Graham T. Allison, in *Essence of Decision: Explaining the Cuban Missile Crisis* (Boston: Little, Brown, 1971), p. 203, argues that the moral argument against the airstrike "struck a responsive chord in the President."

or countervalue strategy made necessary by fewer but larger weapons and their less accurate means of delivery. The first Single Integrated Operational Plan (SIOP), prepared in 1960, offered the president only one option, the use of all existing American nuclear weapons against the Soviet Union, China, and Eastern Europe. It was expected to kill between 360 and 425 million people.[70] One must wonder whether any president could have brought himself to order this kind of holocaust even in response to a Soviet invasion of Europe.

There is another point to consider. Postwar American leaders were never as convinced as were their German counterparts in World War I of the inevitability of war. From the time of Harry Truman on, American presidents refused to believe that the international situation was as bleak as some of their security advisers maintained. Some presidents, like Harry Truman, expressed dire forebodings in response to particularly threatening events abroad. But their pessimistic moods never lasted very long. All the Cold War presidents preferred to believe that war was avoidable and that the Soviet Union might one day mellow and become more moderate in its foreign policy goals. In the 1950s, these hopes were pinned to Stalin's death and the subsequent anticipated transformations of the Soviet political system. In the 1960s, they rested upon the sobering effect that realization of the true destructiveness of nuclear weapons was expected to have upon Soviet leaders. Even the presidents who sustained the darkest view of Soviet intentions, probably Truman and Kennedy, still looked forward to a time when Soviet–American relations might improve. If war was not inevitable, it followed that relative military advantage, while still something of critical importance, no longer dictated a policy of preventive war. Belief in the possibility of avoiding war actually offered American leaders a justification for deferring consideration of such an extreme move.

It is worth entertaining the hypothesis that the hope that war with the Soviet Union could be avoided was something of a motivated bias. As the presidents who would have had to authorize the use of America's nuclear weapons did not want to believe that this would ever become necessary, they revised their estimates of the probability of war downwards. More intriguing still is the possibility that this bias was at least partially self-

---

70. Rosenberg, "'A Smoking Radiating Ruin,'" pp. 3–38; Desmond Ball, *Targeting for Strategic Deterrence*, p. 10; David Alan Rosenberg, "The Origins of Overkill"; Thomas Powers, "Choosing a Strategy for World War III," *The Atlantic*, November 1982, pp. 82–110.

fulfilling. Motivated by moral-psychological needs and after the development of a Soviet nuclear capability, by political-military needs as well, belief in the possibility of avoiding nuclear war may have helped to maintain the peace. It made policymakers cautious rather than risk-prone and more alert than they might have been otherwise to finding ways of preventing war. Even allowing for the current upswing in nuclear anxiety in response to the foreign and defense policies of the Reagan Administration, there are fewer Americans today who expect a nuclear war between the superpowers than at any time during the height of the Cold War.

The steady decline, until recently, in expectations of war is probably more than anything else attributable to the simple fact that the superpowers have managed to avoid a nuclear war for almost four decades. Success to this point has encouraged expectations that they can continue to do so. The German experience before the First World War was just the reverse. According to most historians of the period, the exaggerated German belief in the inevitability of war helped to bring World War I about. Bethmann Hollweg more or less admitted this in 1918: "Yes, my God," he confided to a colleague, "in a certain sense it was a preventive war": "But when war was hanging over us, when it had to come in two years even more dangerously and more inescapably, and when the generals said, now it is still possible without defeat, but not in two year's time."[71] German leaders, seeking to escape from their security dilemma, ignored Bismarck's admonition that preventive war is like committing suicide from fear of death.

The question arises as to why the Germans, unlike their later American counterparts, succumbed to a fatalistic acceptance of war. The answer can probably be traced to their quite different attitudes toward war *qua* war. For the Germans, war was anxiety-provoking in the sense that it was a leap into the unknown, an event full of uncertainties and not without real human cost. Bethmann Hollweg, as we noted earlier, suspected that it would sweep away more thrones than it would prop up. But the idea of war itself held few horrors for most of the German policymaking elite; many of them actually conceived of it as something glorious, manly, and even spiritually uplifting.[72] Moreover, it was generally assumed that victory would have beneficial consequences for German society by creating a mood of patriotic euphoria that

---

71. Quoted in Wolfgang Steglich, *Die Friedenspolitik der Mittelmächte, 1917–1918* (Wiesbaden: Steiner, 1964), Vol. 1, p. 418.
72. Lebow, *Between Peace and War*, pp. 247–254, reviews the literature on this subject.

would overcome the class and party differences threatening the survival of the Reich.

American political leaders after World War II never had any such illusions. They have always viewed nuclear war as horrible and destructive of all important social and political values. Their very different conception of war provided a strong incentive for them to believe that nuclear war was avoidable. American policymakers may actually have shifted their calculus of strategic cost and gain to make it consonant with their psychological needs. By dwelling on the ways in which Soviet foreign policy might become less aggressive and war in the long run unnecessary, they reduced the payoff associated with an immediate preventive war and increased the rewards of procrastination. In doing this, policymakers reversed the causal links between assessment and decision posited by deterrence theory. Instead of threat assessment dictating policy, the preference for peace encouraged an assessment of the adversary that justified that policy.

The Soviet Union: 1982–1986

The German and American experience with windows may offer some insights into contemporary Soviet–American relations. This of course depends upon the degree to which Soviet leaders are also affected by the kinds of political and moral constraints analyzed here. For decades, Western advocates of hard-line defense and foreign policies have denied that this is so. The men of the Kremlin, they allege, have no concern for human life; they would not shrink from unleashing nuclear war even if it resulted in staggering losses for their own people so long as they believed that the Soviet Union would emerge victorious.[73] Richard Pipes, a prominent spokesman for this point of view, has cited Soviet willingness to accept twenty million dead as the price of victory in World War II as corroborating evidence. Pipes and others assert

---

73. Former Commander of the Strategic Air Command, Thomas S. Power, wrote: "With such grisly tradition and shocking record in the massacre of their own people, the Soviets cannot be expected to let the risk of even millions of Russian lives deter them from starting a nuclear war if they should consider such a war to be in the best interests of the Communist cause." *Design for Survival* (New York: Coward-McCann, 1964), pp. 111–113. Albert Wohlstetter, in his influential article "The Delicate Balance of Terror," *Foreign Affairs*, Vol. 27, No. 2 (January 1959), pp. 211–235, made much the same point. "Russian casualties in World War II," he wrote, "were more than 20 million. Yet Russia recovered extremely well from this catastrophe. There are several quite plausible circumstances in the future when the Russians might be quite confident of being able to limit damage to considerably less than this number—if they make sensible choices and we do not."

that a Soviet "bolt-from-the-blue" strike coupled with effective civil defense measures could keep Soviet casualties well below those sustained in World War II.[74]

Pipes' analogy to World War II does not withstand even the most superficial analysis. Stalin did not deliberately choose to sacrifice twenty million Soviet citizens in that conflict. These deaths occurred in the course of a long struggle against an invasion the Soviet dictator had done his best to avoid. Surrender, the only alternative, was out of the question as barbaric German occupation policies convinced most Russians that they had no choice but to fight back, regardless of the cost. In contrast to World War II, a Soviet leader contemplating a surprise attack against the United States would have the clear option of not launching the attack. Unlike Stalin in June 1941, he would have to choose to sacrifice as the price of victory however many million Soviet citizens an American retaliatory strike was expected to kill. A decision to accept such a cost cannot of course be ruled out but there are a number of reasons why it seems improbable. One of them, and here Pipes' own historical analogy can be used to refute his claim, is the Soviet experience in World War II.[75]

Students of Soviet affairs are in general agreement that the Red Army's triumph over Germany in World War II is one of the most important sources of legitimacy of the Soviet political system.[76] Soviet authorities themselves seem to recognize this political reality. They do their best through films, the schools, and diverse forms of propaganda to keep alive the public memory of that conflict and of the leading role played in it by the communist party. To a great extent they have been successful; old men proudly display their medals and campaign ribbons on Sunday promenades, young couples cus-

---

74. Pipes, "Why the Soviet Union Thinks It Could Fight and Win a Nuclear War"; Leon Gouré, *War Survival in Soviet Strategy: USSR Civil Defense* (Oxford, Ohio: University of Miami Center for Advanced International Studies, 1976); Jones and Thompson, "Central War and Civil Defense"; Daniel Gouré and Gordon H. McCormick, "Soviet Strategic Defense: The Neglected Dimension of the U.S.–Soviet Balance," *Orbis*, Vol. 24, No. 1 (Spring 1980), pp. 103–127. For a critique of this literature, see William H. Kincade, "Repeating History: The Civil Defense Debate Renewed," *International Security*, Vol. 2, No. 3 (Winter 1978), pp. 99–120.
75. A similar argument is made by Lawrence Freedman, *The Evolution of Nuclear Strategy* (New York: St. Martin's Press, 1981), pp. 142–143.
76. See, for example, David Holloway, *The Soviet Union and the Arms Race* (New Haven: Yale University Press, 1983), pp. 13–14; Leonard Schapiro, *The Communist Party of the Soviet Union* (New York: Random House, 1960), pp. 494–505; Seweryn Bialer, *Stalin's Successor: Leadership, Stability, and Change in the Soviet Union* (New York: Cambridge University Press, 1980), pp. 183–184.

tomarily lay a wreath at a war memorial prior to their betrothal, and war museums around the country are a weekend haunt of young and old alike. Defense remains a popular cause in the Soviet Union and, as far as outsiders can tell, the tremendous expenditure the regime devotes to it is not resented by the citizenry. Even in the *samizdat* literature, which is sharply critical of most aspects of Soviet life, few objections have been raised about the size of the Soviet military establishment.[77] Leaders who may themselves have little regard for the intrinsic value of human life would still probably be reluctant to embark upon a course of action that threatened to destroy this important source of regime legitimacy.

Most of the men who rule the Soviet Union today occupied responsible positions during "The Great Patriotic War." Their protegés are mostly veterans of that conflict. Even younger officials are likely to have some personal memories of the war. Many of the principal Soviet leaders of the postwar period have given testimony in their speeches, public statements, and occasionally memoirs of the extent to which their country's victory over Germany continues to give them a great sense of personal pride and satisfaction. In varying degrees, they appear to have internalized as important values the survival and well-being of the Russian people and Soviet state.[78]

Nuclear war would destroy much if not all of the economic and social progress Soviet leaders have worked so hard to achieve. George Kennan argues that Soviet leaders are fully aware of this. There is, he claims, "a consciousness on the part of these men that certain of the things they most deeply care about would not be served by Russia's involvement in another great war."[79] Speeches by Soviet leaders in recent years have increasingly stressed this theme, and there is no good reason to doubt the sincerity of their repeated expressions of horror at the prospect of such a conflict.[80] In

---

77. This point is made by Roy A. Medvedev and Zhores A. Medvedev, "A Nuclear Samizdat on America's Arms Race," *The Nation*, January 16, 1982, p. 49; Matthew Evangelista, "Soviet People Support Arms," *In These Times*, March 31–April 6, 1982, pp. 11–12.
78. Holloway, *Soviet Union and the Arms Race*, pp. 102–104, who cites some examples.
79. George F. Kennan, "Reflections: Breaking the Spell," *The New Yorker*, October 3, 1983, pp. 44–53.
80. Some hawkish analysts allege that Soviet statements professing horror at the consequences of nuclear war are part of an elaborate "disinformation" campaign designed to mislead Western opinion and reconcile it to Soviet strategic prowess. See, for example, Dziak, *Soviet Perceptions of Military Power*, pp. 66–67; Joseph D. Douglass, Jr., "Soviet Disinformation," *Strategic Review*, Vol. 9 (Winter 1981), pp. 16–26; William F. Scott, "Continuity and Change in Soviet Military Organization and Concepts," *Air Force Magazine*, March 1982, pp. 47–48. A compelling critique of this argument is offered by Dan L. Strode and Rebecca V. Strode, "Diplomacy and Defense

his keynote address to the Twenty-Sixth Congress of the Communist Party in February 1981, Leonid Brezhnev declared:

"If you ask any Soviet person—whether a member of the Communist Party or not—what has highlighted our Party's path in recent years the answer will be: It was highlighted above all by the fact that we are managing to preserve peace. And for this people of different ages and occupations give their thanks to the Party from the bottom of their hearts."[81]

If the men of the Kremlin are any different from their American counterparts in this regard it may be only in their ability to envisage more vividly the consequences of a nuclear war by reason of their wartime experience.

Only an extraordinary man, or given the quasi-collective nature of the Soviet leadership, an entire group of extraordinary men, would spurn late in life the values that have been so important to their careers and sense of self-worth in favor of some abstract ideological goal. Brezhnev, Andropov, and now Chernenko, as well as the relatively colorless *apparatchiki* who surround them, strike most observers as cautious if hard-nosed men who are at least as intent on preserving what personal influence and international power they have as they are upon expanding it. Like their American counterparts, Soviet leaders seem inclined to do whatever is in their power to avoid ever having to confront a decision to use nuclear weapons. If this portrayal is accurate, any move on their part towards Armageddon would require an unprecedented sense of threat coupled with a decisional format that facilitated every kind of bolstering technique. It is difficult to envisage any situation that meets these criteria other than a crisis in which vital Soviet interests were threatened and in which it appeared to Soviet leaders that the United States was preparing to attack them.[82]

---

in Soviet National Security Policy," *International Security*, Vol. 8, No. 2 (Fall 1983), pp. 91–116, who explain the growing emphasis in statements of both Soviet political and military leaders on the destructiveness of nuclear war as a reflection of a high level decision to stress foreign policy concerns at the expense of purely military issues in order to encourage Western European opposition to Reagan's military policies.

81. L.I. Brezhnev, *Report of the Central Committee of the Communist Party of the Soviet Union to the 26th Congress of the CPSU* (Washington, D.C.: USSR Embassy Information Department, February 1981), p. 2.

82. John Erikson, "The Soviet View of Deterrence: A General Survey," *Survival*, Vol. 24 (November–December 1982), pp. 242–251, argues, "it is reasonable to infer that the sole contingency which could persuade any Soviet leadership of the 'rationality' of nuclear war in pursuit of policy would be the unassailable, incontrovertible, dire evidence that the United States was about to strike the Soviet Union."

Implications for Deterrence

The German and American cases appear to substantiate the importance of nonmilitary considerations for policymakers contemplating the use of force. The moral and psychological costs of aggression clearly influenced both sets of leaders. For the Germans, it acted as a barrier to war in the absence of a decisional context that permitted them to deny to themselves their responsibility for the conflict. For the Americans, it was and still is an important constraining factor.

The absolute as opposed to the relative cost of war can also function as an important constraint to war. This was most apparent in the American case where the available evidence indicates that presidents would be unwilling to accept the loss of one or more American cities even if it would permit them to destroy their principal adversary. This indicates that, as the absolute cost of war increases, the importance of the relative gains diminishes and may ultimately become irrelevant to the decision for war or peace. "Self-deterrence," brought about by the anticipated domestic political, moral, and psychological costs of a foreign policy, may be at least as important a motive for moderation as deterrence in the traditional sense, brought about that is by fear of external punishment. Deterrence theorists have generally failed to take these considerations into account.

Deterrence theory should not be condemned because of the faults of its practitioners. There is no reason why it could not attempt to incorporate the political, economic, and psychological values of peace as important components of a leader's calculus of cost and gain.[83] Passivity in the face of an obvious opportunity to take advantage of an adversary could be made consistent with the assumption of the theory by invoking these nonmilitary costs and demonstrating how they exceeded the expected utility of the action. Arguments of this kind would, of course, risk making deterrence theory tautological as they would likely be invoked as a *deus ex machina* to explain away moderation in circumstances in which more aggressive behavior would

83. The author is indebted to James E. King for this idea. In conversations over the years he has frequently criticized deterrence theorists for limiting their horizons to military considerations. One early and limited attempt to go beyond this in the context of a model of the process by which the Soviet Union might decide to launch a first strike was Daniel Ellsberg, "The Crude Analysis of Strategic Choices" (Santa Monica, Calif.: The Rand Corporation, 1960). Ellsberg included some political variables in his equation but weighted them quite arbitrarily and gave no inkling whatsoever as to how they might be measured.

have been expected. It is also difficult to imagine how, in practice, political, moral, and psychological values could be factored in. They are intangible considerations that do not lend themselves to ready assessment let alone to quasi-quantification. Some precision in measurement would nevertheless be essential for any attempt to rank the relative importance of an adversary's values. Only then could the trade-offs be calculated between aggression and passivity. The results of this calculation would also have to be specified before the outcome became known in order to keep the theory from becoming a tautology.

The complexity, if not the practical impossibility, of measuring and comparing the relative importance of diverse and by no means necessarily compatible values is probably an important reason why most deterrence theorists have tended to restrict their horizon to relative military capability. A unidimensional perspective finesses altogether the thorny problem of value comparisons. As there is only one value to be considered, military prowess, it becomes the absolute and unquestioned benchmark for assessing deterrence. If the requisite military capability is lacking—and how that is determined is another matter—then deterrence is found wanting.[84]

Relative military capability has the further appeal that it appears to lend itself to quantification more readily than do other possible measures of deterrence. Relative military strength is frequently described in terms of the respective makeup and size of adversarial forces and the performance characteristics of their most important weapons systems. In the strategic realm, these comparisons can be static or dynamic, the latter based on models of force exchanges under varying conditions.[85] However, these calculations and comparisons are often misleading because they ignore qualitative considerations of leadership, morale, strategy, and doctrine, which the history of warfare indicates are collectively more important than numbers of men and weapons in determining the outcome of conflicts.

---

84. For a recent example of a study that assesses deterrence only in terms of military considerations, see John Mearsheimer, *Conventional Deterrence* (Ithaca: Cornell University Press, 1983). His justification is "the difficulty, if not impossibility, of developing a theory that takes these [non-military] elements into account" (p. 14).

85. The best discussion of static measures is in Thomas A. Brown, "Number Mysticism, Rationality, and the Strategic Balance," *Orbis*, Vol. 21, No. 3 (Fall 1977), pp. 479–490. For the problems inherent in dynamic measures, see Lebow, "Misconceptions in American Strategic Assessment," which discusses models used by the Defense Department. William H. Baugh, *The Problems of Nuclear Balance* (New York: Longman, 1984), pp. 122–155, describes several dynamic models of strategic exchanges.

If comparisons of military capability are theoretically misleading, they are also politically dangerous. They encourage exaggerated perceptions of threat to the extent that they rely on worst case analysis, an evil seemingly endemic to force comparisons and strategic exchange models devised by professional military and civilian analysts. As a general rule, these analysts are most sensitive to the capabilities of their adversary and the deficiencies of their own forces. They must also base their models on uncertain and incomplete data about the performance characteristics and operational reliability of weapons on both sides, but especially those of the adversary. The less that is known about the qualities of the other side's weapons, the greater the tendency to assign high values to them in order to be on the "safe side." In dynamic analyses, this bias can be further compounded by the choice of a war scenario that is particularly favorable to the enemy. In a strategic exchange this is likely to be a "bolt-from-the-blue" strike at a time when one's own forces are "ungenerated," that is in a day-to-day state of readiness. Rigging the situation in this way results in an extremely threatening picture of the strategic balance.[86]

When worst case analysis is used by both sides, it means that they will interpret a situation of strategic parity as one of imbalance favoring their adversary. This will encourage both states to augment or modernize their arsenals in order to redress the balance. This in turn will aggravate the tensions between them as each side will interpret any arms buildup as proof of the other's hostile intentions given its belief that its adversary already possesses an advantage. The current Soviet and American inability to agree about either the conventional or nuclear balance in Europe offers a telling example of just how this dynamic operates.[87] It illustrates how asymmetrical

86. Sienkiewicz, "Observations on the Impact of Uncertainty in Strategic Analysis," shows how even small variations in the numbers built into these models can result in significantly different outcomes. He calculated that for every 5 percent of the Minuteman force that survives a Soviet first strike the United States retains the capability to destroy 300 Soviet targets in a retaliatory strike. Needless to say, none of the dynamic analyses used to assess the strategic balance can pretend to predict the percentage of Minutemen surviving a Soviet strike with anything approaching a 5 percent level of confidence.
87. For a documentation of this phenomenon see the twin articles by Raymond L. Garthoff, "The Soviet SS-20 Decision," *Survival*, Vol. 25 (May–June 1983), pp. 110–119, and "The NATO Decision on Theater Nuclear Forces," *Political Science Quarterly*, Vol. 98 (Summer 1983), pp. 197–214; Gert Krell, *Deterrence in the 1980's: Part X, The Limitations of Nuclear Deterrence: Criteria For Restraint*, Adelphi Paper (London: International Institute for Strategic Studies, forthcoming); Jane M.O. Sharp, "Is European Security Negotiable?," in Derek Leebaert, ed., *European Security: Prospects for the 1980's* (Lexington, Mass.: Lexington Books, 1979), pp. 261–296.

perceptions of military balance are an important structural cause of arms races.

When perceptions of imbalance are wedded to fears of windows of vulnerability, threat perception becomes more exaggerated still. For now, the adversary is seen not only to have the wherewithal to carry out an effective attack but also the incentive to do so. Such an analysis, we have argued, is intellectually naive. It is also a poor predictor of state behavior. Germany went to war in 1914 even though its perceived window of opportunity was more pronounced in 1905, 1909, and 1912. To the extent that German leaders were influenced at all by relative military capabilities, it was by the fact that Germany's advantage was seen to be rapidly *diminishing* in 1914; they went to war to forestall a window of vulnerability. And even that, we have tried to show, was an insufficient condition for war. Germany's leaders also required a decisional context that permitted them to deny all responsibility of war to themselves. The United States failed to exploit its window in the 1950s and early 1960s. So did the Soviet Union *vis à vis* China a decade later.

If states often fail to exploit windows of opportunity, they also start wars at inauspicious moments judged in terms of relative military capabilities.[88] World War II is a case in point. Hitler, a leader certainly unconstrained by the usual political and moral considerations, failed to time his war to correspond with the period of Germany's maximum expected military advantage. German military planners predicted that this would not be until about 1943. Hitler chose war in 1939 and there is evidence that he would have actually preferred it in 1938 when Germany was even less prepared. Hitler had a personal timetable. His obsession with his health, fear of a premature death, and belief that only he could lead Germany to victory drove him to provoke war while he was fifty and still healthy.[89] The Argentine invasion of the Falklands provides a more recent example. Had the Argentines postponed their attack one more year the British would have decommissioned so many of the ships vital to their invasion force, among them the aircraft carrier *Hermes*, that an effort to retake the Falklands would probably no longer have been considered a viable policy option. The Argentine *junta* did not wait

88. For a fuller treatment and documentation of this assertion, see Lebow, "Conclusions," in Jervis, Lebow, and Stein, *Psychology and Deterrence*.
89. Gerhard Weinberg, *The Foreign Policy of Hitler's Germany*, 2 vols. (Chicago: University of Chicago Press, 1970–1980), Vol. 2, p. 663.

because their action was taken in response to internal political need, not external military opportunity.[90]

As war is an extension of politics by other means, its objectives and timing are generally determined by *political* considerations. Attempts to predict war on the basis of the military balance are therefore likely to be misleading. In the first instance, they will encourage predictions of wars that never come to pass. They will also make analysts insensitive to the prospects of war in situations where the military balance is not favorable to the would-be aggressor. The Israeli intelligence failure in October 1973 has been attributed to this latter phenomenon.[91]

There is another danger to window of opportunity analysis. When leaders believe in windows they risk making themselves vulnerable to them. If an adversary can gain a military advantage, or merely convince the other side that it possesses one, it can exploit its putative advantage for political ends. Hitler did this in the 1930s. He succeeded in portraying German military might, especially airpower, as much greater than it was at the time. Fear of German power overlaid on French and British expectations that Hitler would not hesitate to resort to war to achieve his goals was one of the principal causes of appeasement.[92]

To the degree that American leaders are convinced that the Soviet Union would start a war simply because the "correlation of forces" was favorable to it, they open themselves up to political blackmail should Soviet leaders ever succeed in convincing the United States that they possess a significant strategic advantage. Given the baneful effects of worst case analysis, Moscow

---

90. Richard Ned Lebow, "Miscalculation in the South Atlantic: The Origins of the Falkland War," *Journal of Strategic Studies*, Vol. 6 (March 1983), pp. 5–35.

91. The Agranat Report, *A Partial Report by the Commission in Inquiry to the Government of Israel* (Jerusalem: Government Press Office, April 2, 1974). Israel's leaders operated on the flawed assumption that Egypt would only go to war together with Syria to seek Israel's destruction. It followed that Egypt would not attack unless two conditions were met: the Egyptian air force had to be capable of striking at Israel in depth, in particular at Israeli airfields, and Egypt had to be joined in the attack by Syria. Israeli military intelligence accordingly considered an Egyptian attack unlikely before 1975, the earliest possible date they believed the Egyptian air force capable of absorbing enough Soviet equipment to achieve the requisite strike capability. For the best secondary treatment of this case, see Janice Gross Stein's two studies that treat it from both the Israeli and Egyptian perspectives: "'Intelligence' and 'Stupidity' Reconsidered: Estimation and Decision in Israel, 1973,'" *The Journal of Strategic Studies*, Vol. 3 (September 1980), pp. 147–177, and "Calculation, Miscalculation and Conventional Deterrence: The View from Cairo," in Jervis, Lebow, and Stein, *Psychology and Deterrence*.

92. Uri Bialer, *The Shadow of the Bomber: The Fear of Air Attack and British Politics, 1932–39* (London: Royal Historical Society, 1980). The wider implications of the case are discussed by Robert Jervis, "Deterrence and Perception," *International Security*, Vol. 7, No. 3 (Winter 1982–83), pp. 3–30.

would not really need much of an advantage to attempt to convince Washington of its superior prowess. Fortunately, the Soviet Union has for the time being chosen to downplay rather than exaggerate its strategic capability.[93] This does not detract from the political truth that the kinds of window of vulnerability scenarios that have been widely publicized in recent years do a serious disservice to the real interests of the United States to the extent that they are taken seriously by policymakers.

---

93. Marshal Rodion Malinovsky adumbrated in January 1962 the policy line that the Soviet Union has more or less adhered to ever since. He declared: "I hold that today the socialist camp is stronger than these countries [NATO], but let us presume that the forces are equal. We are ready to agree to this so as not to take part in stirring up a war psychosis. But since the forces are equal the American leaders should come to correct conclusions and pursue a reasonable policy." Arnold Horelick and Myron Rush, *Strategic Power and Soviet Foreign Policy* (Chicago: University of Chicago Press, 1966), p. 88. Holloway, *Soviet Union and the Arms Race,* pp. 70–80, refers to later speeches of Brezhnev and Ogarkov that make the same point. Andrew Cockburn, *The Threat: Inside the Soviet Military Machine* (New York: Random House, 1983), pp. 275–276, tells the story of Chairman of the Joint Chiefs Nathan F. Twining's trip to the Soviet Union in 1956. He was told by Defense Minister Zhukov that the Americans were routinely "too high in estimating our strength." Twining and other members of the delegation were convinced that this was an egregious piece of disinformation.